# Multimedia
# Messaging Service

# Multimedia Messaging Service

## An Engineering Approach to MMS

**Gwenaël Le Bodic**

*Alcatel, France*

John Wiley & Sons, Ltd

This publication is designed to provide accurate and authoritative information in regard to the subject
matter covered. It is sold on the understanding that the Publisher is not engaged in rendering
professional services. If professional advice or other expert assistance is required, the services of a
competent professional should be sought.

***Other Wiley Editorial Offices***

John Wiley & Sons Inc., 111 River Street, Hoboken, NJ 07030, USA

Jossey-Bass, 989 Market Street, San Francisco, CA 94103-1741, USA

Wiley-VCH Verlag GmbH, Boschstr. 12, D-69469 Weinheim, Germany

John Wiley & Sons Australia Ltd, 33 Park Road, Milton, Queensland 4064, Australia

John Wiley & Sons (Asia) Pte Ltd, 2 Clementi Loop #02-01, Jin Xing Distripark, Singapore 129809

John Wiley & Sons Canada Ltd, 22 Worcester Road, Etobicoke, Ontario, Canada M9W 1L1

Wiley also publishes its books in a variety of electronic formats. Some content that appears
in print may not be available in electronic books.

***British Library Cataloguing in Publication Data***

A catalogue record for this book is available from the British Library

ISBN 0-470-86253-X

Typeset in 10.5/13pt Times by Laserwords Private Limited, Chennai, India
Printed and bound in Great Britain by TJ International, Padstow, Cornwall
This book is printed on acid-free paper responsibly manufactured from sustainable forestry
in which at least two trees are planted for each one used for paper production.

# Contents

*The nice thing about standards is that there are so many to choose from*

*Andrew S. Tanenbaum*

# Preface

First wireless communications over radio links were demonstrated by Guglielmo Marconi in the 1890s. These initial communications trials opened the door to many applications from radio to television broadcasts. The first regular radio broadcast service began in 1920 and the commercial introduction of colour television followed more than 30 years later. Colour television is a significant milestone in the road map of radio-based broadcast services since it allowed voice to be transmitted along with motion colour pictures.

Radio-based communications also allowed the introduction of another category of innovative applications: personal communications between mobile users. To some extent, personal communications follow a similar evolution path as that of commercial radio-based broadcast services. Commercial personal mobile communications services started as voice-centric services with the introduction of the first radio mobile network in 1981 in Europe. Twenty years later, advances in high technology made possible multimedia personal communications over radio links between mobile users. The Multimedia Messaging Service (MMS) is one of these applications allowing users to exchange multimedia messages. A multimedia message can be a simple text or voice message or a sophisticated message containing various media objects (e.g. video clip, image, melodies). Regarding the importance of MMS in the evolution of personal communications, MMS may soon represent for radio-based personal mobile communications services what colour television is to radio-based broadcast services.

The first reference to the term 'Multimedia Messaging Service' can be traced back to 1998. At this crucial time, SMS usage was booming and new requirements for enhancing the messaging experience of mobile users were emerging. To meet this demand, major industry players defined the scope of what was expected to become a ubiquitous multimedia messaging service with the intention to deploy it over various transport technologies for mobile networks.

MMS can be seen as the 'best of the breed' of several messaging services including the well-known Short Message Service (SMS) and the Internet electronic mail. Year 2002 saw the first wave of MMS devices appearing on the market, mainly targeting photo messaging with the availability of camera phones. Year 2003 sees the emergence of a new generation of MMS devices with video capabilities. MMS is still in its infancy, and ongoing standardization developments should allow further MMS use cases. As for any new service launch in a multi-vendor environment, current MMS solutions are still subject to interoperability issues, preventing a wide acceptance of the service by the mass market. Vendors are currently addressing these issues.

Two international organizations, the 3rd Generation Partnership Project and the WAP Forum, took the challenge of designing the initial MMS standards, allowing the development of interoperable devices. The Open Mobile Alliance and the 3rd Generation Partnership Project 2 are two other standardization organizations that recently took the responsibility to publish complementary MMS standards.

The core of this book consists of a comprehensive description of enabling technologies for MMS, illustrating as much as possible standards published by international organizations. This book also provides an introduction to market needs for MMS and compares MMS features with the ones of other messaging services such as SMS, electronic mail or Japan's Sha-Mail and i-mail.

Chapter 1 provides a general introduction to MMS, identifying major use cases and comparing MMS with other messaging services. Chapter 2 demystifies the working procedures of standardization organizations and explains how standards are referenced and versioned. This chapter will prove to be a valuable material for engineers who have to manipulate standards. Chapter 3 outlines the elements composing the MMS architecture and introduces the required communications interfaces. The Wireless Application Framework is also introduced in this chapter. Chapter 4 describes each feature offered by MMS, from a usage perspective. This description encompasses the sending and retrieval of messages, the management of reports, methods for setting MMS devices and so on. The multimedia message is a key element for MMS. Chapter 5 is dedicated to the presentation of the message structures and identifies the various media objects (text, images, video clips, scene descriptions, etc.) that can be included in a message. Chapter 6 offers an in-depth description of standardized transport protocols for several of the available MMS communications interfaces. Chapter 7 presents standard conformance and interoperability testing aspects. Chapter 8 outlines commercial solutions available on the market for MMS. This includes MMS phones, MMS centres and developer tools. Chapter 9 provides an insight on how MMS could evolve in the future. Last but not least, the Appendix section of this book groups a number of useful information such as commonly used content types and interface error codes. This section is intended to be an easily accessible concise reference for those involved in the development of MMS software applications.

This book sometimes builds up from selected materials published as part of the author's previous book, covering general mobile messaging aspects: *Mobile Messaging* (Wiley & Sons, Ltd, November 2002). In particular, Chapter 2 that deals with standardization aspects is built up from a similar standardization chapter of *Mobile Messaging*. However, Chapter 2 has been entirely revised. In this chapter, parts of the description of the WAP Forum working processes have been removed since the WAP Forum has now been merged with other groups to form the Open Mobile Alliance. A full description of working procedures for the Open Mobile Alliance has consequently been introduced. Chapter 6 provides an in-depth description of transaction flows between elements composing the MMS environment. Content of Chapter 6 is also partly derived from *Mobile Messaging*. However, this book provides easier-to-read parameter tables, clear graphical XML representation of data structures, and has been extended to cover the most recent MMS features such as multimedia message boxes.

The author would like to gratefully acknowledge the time and effort of many people who reviewed the content of this book. The book has benefited from constructive comments from experts involved in various MMS activities (standardization bodies, mobile network operators, handset manufacturers and third-party application developers). In particular, the author is thankful to Eskil Åhlin, Stéphane Augui, Philippe Delaloy, Thibaud Mienville, Pierre Grenaille, Michael Ishizue, Marie-Amélie Le Bodic, Thomas Picard, Friedhelm Rodermund and Andreas Schmidt.

The team at John Wiley & Sons Ltd involved in the production of this book, provided excellent support and guidance. Particularly, the author is grateful to Daniel Gill, Mark Hammond and Sarah Hinton for their continuous support during the entire process.

The bibliography lists a number of standards that are useful for exploring further topics introduced in this book. Pointers to these standards and other useful resources are available from this book's companion website at
http://www.lebodic.net/mms_resources.htm

Gwenaël Le Bodic, Ph.D.

## Acknowledgements

Images of mobile phones have been used courtesy of Alcatel Business Systems, Orange SA, Siemens AG and Sony Ericsson.

# About the Author

Gwenaël Le Bodic is a mobile Internet and standardization expert for Alcatel's mobile phone division (France). His activities for Alcatel include participating and contributing to the development of messaging technologies and services in the scope of the 3GPP and OMA standardization processes. He has been responsible for the design of the software architecture of the messaging solution for Alcatel's first two MMS phones. A certified engineer in computer sciences, Gwenaël Le Bodic obtained a Ph.D. in mobile communications from the University of Strathclyde, Glasgow. He is the author of many research publications in the field of mobile communications. He wrote the book *Mobile Messaging* (Wiley & Sons Ltd, November 2002) covering various messaging services and technologies. Gwenaël Le Bodic can be contacted at gwenael@lebodic.net.

# 1

# Introduction to MMS

The Multimedia Messaging Service (MMS) can be seen as the 'best of the breed' of several messaging services such as the Short Message Service (SMS), the Enhanced Messaging Service (EMS) and the Internet mail. The first SMS short text message is believed to have been exchanged in 1992. Ten years later, MMS multimedia messages propagate over radio channels of major mobile networks. In between, EMS, designed as a rich media extension of SMS, attempted to penetrate the market but without great success. Since 2002, a first MMS wave has spread all over the world, with more than 100 operators adopting the service. The first MMS wave offers basic messaging features to mobile users and a second MMS wave is already appearing. This second wave builds up from basic messaging functions to offer more sophisticated features, from photo messaging to video messaging. In 2003, MMS is still in its infancy and still has to meet the expectations of the mass market.

From 2002, the first MMS wave has led to the roll-out of the service in many countries in Europe, Asia and North America. This first market opportunity for MMS relied mainly on the availability of colour-screen phones with digital camera and the introduction of packet-based communications in mobile networks. This first wave of MMS allows mobile users to exchange multimedia messages with the Internet and mobile domains. Multimedia messages range from simple text messages to sophisticated messages comprising a slideshow composed of text, images and audio clips. The roots of the Multimedia Messaging Service lie in the text-based Short Message Service and the Internet electronic mail. Indeed, features already supported by these services have not been forgotten in MMS. MMS supports the management of reports (delivery and read reports), message classes and priorities and group sending. In addition, MMS differs from other messaging services with its multimedia capabilities, its support for email and phone number addressing modes, its efficient transport mechanism and flexible charging framework. From a marketing perspective, the first MMS wave is mainly regarded as the 'photo messaging' service for the mass market of mobile users. It is too early to speak about an overwhelming commercial success for the first MMS wave. Operators have widely adopted the service but interoperability issues are still to be solved and penetration of MMS phones has to grow in order to allow a mass adoption of the service by mobile subscribers.

*Multimedia Messaging Service: An Engineering Approach to MMS*   Gwenaël Le Bodic
© 2003 John Wiley & Sons, Ltd   ISBN: 0-470-86253-X

As the first MMS wave was crossing the globe, open standards for MMS were evolving to enable future service evolutions and solving early interoperability issues. These evolutions represent the basis of the emerging second MMS wave, likely to start by the end of 2003. It will leverage the first MMS wave with the support of new features and new media formats. Certain MMS solutions already support the exchange of larger objects such as video clips. This will progressively lead to the transport and storage of larger messages. In the context of MMS, the concept of multimedia message box (MMBox) will ease the management of large messages by allowing the storage of multimedia messages in network-based user personal stores (e.g. message boxes, online photo albums, etc.). The wide-scale deployment of these new features is still to be accomplished by mobile operators. The support of the second MMS wave faces interesting technical and marketing challenges. This book attempts to demystify those.

This chapter places MMS in the patchwork of existing messaging services. It identifies the key success enablers for MMS and compares MMS features with those offered by other messaging services.

## 1.1 MMS Success Enablers

The commercial introduction of MMS started in March 2002. The future success of MMS is believed to rely on four main enablers:

- *Availability and penetration of MMS phones*: Mobile users require MMS-enabled phones for composing and sending multimedia messages. Availability of phones is less critical for message reception and viewing since, with message transcoding in the network side, users are often able to send messages to Internet users (via email) and to users of legacy handsets (non-MMS phones with support of SMS and/or WAP browser). However, a certain market penetration of MMS-enabled phones is required to enable significant revenues. The Global Mobile Suppliers Association[1] believes that a penetration of at least 30% is necessary for MMS to succeed, and it expects this level of penetration to be reached by end of 2003. In 2002, MMS started with a very limited number of MMS phones. At the time of writing, more than 50 MMS phone models (see Chapter 8) were available, and this figure is increasing at an impressive rate. MMS phones require the support of colour screens and are often shipped with a built-in digital camera. Obviously, these multimedia phones are relatively expensive to produce but mobile operators are ready to strongly subsidize the cost of producing phones in order to facilitate a rapid roll-out of the service. The mass production of MMS-enabled phones will lead to an economy of scale, and this will further increase the market penetration of these devices.
- *Device interoperability and service interworking*: The introduction of any new telecommunications service in a multi-vendor environment is always subject to

---

[1] http://www.gsacom.com/

equipment interoperability issues. Such an interoperability issue occurs, for instance, when two vendors of communicating devices interpret a standard differently. In the context of MMS, the number of standards and the number of vendors offering solutions are high; therefore the interoperability risk is proportionally high. Although the MMS standards have been designed with greatest care, too many options sometimes lead to the development of devices conforming to the standard, but which do not interoperate in an efficient manner. Initially, service interworking between MMS providers (typically mobile network operators) was seldom ensured. This made the exchange of multimedia messages among subscribers belonging to different MMS domains complicated. Lack of service interworking was mainly due to the non-existence of commercial agreements between MMS providers. These agreements are being negotiated now, and service interoperability barriers between MMS providers are being removed.

- *Ease of use*: Snapshot and send! The use of MMS should be as easy as this. No time for browsing through complex phone menu items. The use of MMS with the phone should be facilitated with dedicated buttons and simplified options, and message sending should be realized with a minimum of track point clicks. Besides the man-machine-interface issues, another cornerstone to achieve ease of use is the availability of pre-configuration methods for MMS settings. This encompasses the storage of default MMS settings during the device manufacturing process, the storage of settings in the SIM card or the provisioning of settings over the air (e.g. settings are sent dynamically from the network to the device).

- *Added value for the end-user*: The user should perceive significant added value using MMS compared to other messaging systems such as SMS or email. Added value of MMS includes its multimedia capabilities, an efficient message transport mechanism, the support of various addressing modes and management of reports (e.g. delivery and read reports). Added value is also provided by enabling mobile users to enjoy new types of information, entertainment and other services.

MMS is in its infancy. At present, much hype surrounds MMS, but it still has to prove that it can fulfil the four success enablers as described above. MMS has the key advantage of having full support from the major players of the mobile communications industry. Indeed, in a mobile phone market where the penetration rate is high, MMS is an opportunity for device manufacturers to replace the legacy voice-centric phones by selling new sophisticated multimedia phones. Operators regard MMS as the revenue-generating service that is appropriately scaled for recent investments in terms of packet-based transport technologies (e.g. GPRS) leading to a smooth transition to the forthcoming roll-out of 3G networks. MMS bridges the once closed mobile communications world with the Internet domain, opening the door to the deployment of compelling services by innovative Value-Added Service (VAS) providers. Without any doubt, the entire industry has great expectations for the future of MMS. The future will tell if the actual hype will convert into commercial success.

## 1.2 Commercial Availability of MMS

Telenor from Norway was the first operator to launch MMS in Europe in March 2002. This initiative was followed by Vodafone D2 (April 2002), Westel Hungary (April 2002), Telecom Italia Mobile (May 2002), Orange UK (May 2002), Swisscom (June 2002), Orange France (August 2002), T-Mobile Germany/Austria (summer 2002), T-Mobile UK (June 2002), Vodafone UK (summer 2002), Telefonica Moviles Spain (September 2002) and others.

Outside Europe, China Hong Kong CSL launched MMS in March 2002 and was followed, shortly afterwards, by other local operators. In the United States, AT&T Wireless launched MMS in June 2002. In Singapore, Singtel Mobile launched MMS in September 2002 and China Beijing Mobile launched MMS in China in October 2002.

In the first quarter of 2003, more than 100 operators around the world have announced the availability of their MMS services. The service is now available worldwide and MMS is gaining thousands of new users every day.

## 1.3 MMS Compared with Other Messaging Services

The first usage of the term 'MMS' dates back to 1998. At that time, operators and vendors were looking at opportunities to offer a messaging service for third-generation mobile systems. Considering the success of SMS, standardization work on MMS was rapidly kicked off. In this context, MMS can be considered as the 'best of the breed' of several existing messaging services. This section describes several messaging services that are close to MMS in terms of underlying concepts and offered features.

### 1.3.1 SMS and EMS

The roots of mobile messaging in Europe lie in the Short Message Service. In its initial form, SMS is a basic service for exchanging short text messages (with a maximum of 160 simple characters). The first text message is believed to have been transferred in 1992 over signalling channels of one of the major European GSM networks. Since this successful trial, SMS usage has been the subject of a tremendous growth reaching 1.5 billion SMS messages sent across the United Kingdom's four GSM networks in February 2003 (source: Mobile Data Association[2]). Despite its limitations, SMS is widely used today and accounts for a significant part of mobile operator revenues. In its most recent form, SMS allows short text messages to be concatenated to form larger messages, and several application-level extensions have been designed on top of SMS as a transport technology. Most notably, EMS is a standardized extension allowing SMS messages to incorporate rich media such as polyphonic melodies, simple black and white, colour or greyscale images/animations and so on. Major phone manufacturers such as Alcatel, Motorola, Siemens and Sony-Ericsson have released EMS-enabled phones. Another application-level extension of

---

[2] http://www.mda-mobiledata.org/

SMS is known as 'Picture Messaging' (part of Smart Messaging services). Picture messaging is a proprietary service developed by Nokia and available mainly on Nokia phones. Features offered by picture messaging are similar to the ones offered by EMS. Unfortunately, the two services have not been designed to interoperate.

SMS was originally developed as part of the GSM technical specifications from ETSI. SMS standardization work was later transferred to the Third Generation Partnership Project (3GPP).

### 1.3.2 Electronic Mail

One of the most common uses of the Internet is the electronic mail (email). First email systems were very basic and cumbersome but were quickly improved with the support of group sending, message attachments, automatic message forward and so on. Email has now become the universal messaging service for Internet users. In the past, email used to be limited to the exchange of plain text messages, sometimes with binary attachments. Now, the text part of email messages can be formatted with HTML, allowing more sophisticated message presentations (inline images, tables, formatted text, etc.).

An email user usually has an email service subscription with a service provider (Internet service provider or other). The email architecture is typically based on an interconnection of local email clients and email servers. The email client is used for the composition and sending of messages to the email server. It is also used for retrieving messages from the email server. The email server is responsible for storing messages in user mailboxes and is often interconnected with other email servers to allow the exchange of messages between distinct email systems.

The email client is typically in charge of retrieving messages from the email server without explicit notification of message availability from the email server. Retrieval of messages can be triggered explicitly by the email user, or the email client can automatically poll the email server for messages awaiting retrieval. This polling mechanism is not appropriate for mobile radio systems, which still have very limited network bandwidth compared to fixed networks. Furthermore, the size of email messages can reach several megabytes. Today, such large message sizes are still difficult to manage with mobile systems. Several phone vendors have attempted to ship devices with embedded email clients but these attempts have not proved to be very successful. Email extensions have been developed to cope with the limitations of mobile systems. One of the successful proprietary extensions of the email system is commercially available in the form of the Blackberry service as described later in this chapter.

### 1.3.3 J-phone's Sha-mail and NTT Docomo's i-shot

In November 2000, J-Phone, the Japanese arm of Vodafone, launched a new messaging service known as 'Sha-mail' (literally stands for 'Picture mail' in Japanese). In October 2002, Vodafone reported that Japan's J-Phone had 7 million Sha-mail handsets operating on its network. Sha-mail is a messaging service for taking photos with a digital camera built into a mobile phone and sending them to another Sha-mail phone or to an Internet user (electronic mail message with picture as an attachment). A service extension of Sha-mail, known as 'Movie Sha-mail', also allows recording and sending short video clips (up to 5 seconds). Sha-mail messages can be stored in Sha-mail digital albums stored in the network and managed remotely by the user via a Sha-mail phone. With Sha-mail, there is no application or monthly fee and customers are only billed for communication charges (based on volume of data).

NTT Docomo is well known for its successful i-mode services launched in February 1999 in Japan. In December 2002, NTT Docomo claimed that 36 million i-mode users have been provided access to the service. The denomination 'i-mode' refers not only to a technology for accessing the Internet from a mobile phone but also to the entire i-mode value chain including technologies, business model and marketing. i-mode offers services such as browsing (access to Internet sites with i-mode-tailored contents), downloading (ringtones, Java applications, etc.) and messaging. i-mode messaging, also known as 'i-mail', is basically based on the Internet electronic mail technology, as described in the previous section. The success of i-mode has spread to other countries outside Japan. Several operators have introduced i-mode in Europe (E-plus of Germany, KPN of Netherlands, BASE of Belgium and Bouygues Telecom of France) and Taiwan (KG Telecom). In response to the success of J-Phone's Sha-mail, NTT Docomo counter-attacked with the launch of a new i-mode messaging service known as 'i-shot'. With i-shot, users can take photos with an i-mode phone with a built-in camera. The photo is attached to an electronic mail message (JPEG file up to 30 kB) and sent to the i-shot server. The i-shot server stores the photo and sends a URL referring to it as part of an email text message to the recipient(s). During this process, the i-mode server may modify the original photo according to the recipient's i-mode device capabilities. Upon reception of the message, the user reads the text message with the i-mode mail client and can directly launch the browser to fetch the photo identified by the URL. The i-shot service is also open to the Internet. In this context, the message is directly transferred to the recipient Internet user as an email message with the photo as an attachment. A key advantage of i-shot is that i-shot messages can be fetched and viewed from any i-mode phone shipped with an i-mode browser. An i-shot phone is only required for originating an i-shot message. With i-shot, there is a monthly fee for accessing i-mode services, and customers pay for communication charges (based on volume of data).

J-phone's Sha-mail and NTT Docomo's i-shot are messaging services for second-generation mobile systems targeted at the mass market of mobile customers. They are proprietary services relying on existing Internet-based protocols and controlled by operators (NTT Docomo, J-phone and other operator partners). At the time of writing, no third party was known to offer Sha-mail or i-shot services. Both services are open to the Internet.

The success of photo messaging services in Japan seems quite encouraging for the success of MMS in other parts of the world. However, Japan is a more data-driven market with shorter handset-replacement cycles, and, therefore, one cannot transfer the Japanese experience directly to other markets.

### 1.3.4 RIM's Blackberry

In the context of mobile communications, it was shown earlier that Internet electronic mail solutions have proven to be very impractical to use without a minimum of adaptation to the constraints of mobile devices and networks. The major barriers to the success of these solutions are the 'pull' model for retrieving messages, which requires frequent accesses to the email server and the fact that server access protocols are not bandwidth efficient. In order to offer an Internet electronic mail service scaled to the requirements of mobile subscribers, the Canadian company Research in Motion (RIM) designed a set of extensions for the existing Internet email service. This extended service, offered to subscribers under the denomination 'Blackberry service', bypasses email inadequacies to the mobile domain by enabling

- a 'push' model for message retrieval,
- compression of messages,
- an encryption of messages.

Two main configurations are available for the Blackberry service. The first configuration limits the impact on existing email architectures by integrating a 'desktop' Blackberry application (the Blackberry desktop redirector) in the user's personal computer used for accessing email messages. When the user is on the move, the desktop application intercepts incoming messages, compresses them, encrypts them and pushes them to the Blackberry device via a mobile network. The other way round, the user can compose a new message with the Blackberry device. The message is compressed and encrypted by the device and sent via the mobile network to the desktop application. The desktop application receives the message (by polling the email server), decompresses and decrypts it and sends it normally to the message recipients as if the message had been sent out directly by the user from his/her personal computer. A more sophisticated configuration of the Blackberry service consists of installing an extension to the email server itself (the Blackberry enterprise server). In the second configuration, the user's personal computer does not have to be left running when the user is on the move. With this configuration, messaging functions performed by the desktop application in the first configuration are performed here by the server extension. In addition, this configuration also allows the synchronization of calendaring and scheduling data between shared corporate databases and remote Blackberry devices.

The Blackberry service first started in North America and has now been deployed in other countries in Europe (e.g. United Kingdom and France). The service fulfils particularly well the needs of itinerant professional users, who avoid using laptop computers while on the move (because of long dial-up time for accessing email servers, etc.). Compared with the other messaging services described in this section, the Blackberry service targets professional users rather than the mass market of mobile users.

## 1.4 MMS Added Value and Success Factors

Why design a new messaging service in the form of MMS when there are so many existing services to choose from? In the late 1990s, SMS usage was booming and major mobile market players were looking for new service opportunities to exploit network resources for the coming years. It was understood that SMS was very limited and mobile messaging services had great margins for improvement. The Internet electronic mail available at this time was not optimized enough for low-bandwidth radio networks and input-limited mobile devices. Japanese photo messaging services were under development in a proprietary fashion and therefore could not meet the market demands in all parts of the world. What was then needed was a universal messaging service offering multimedia features to the mass market of mobile users. MMS builds up from SMS, email and emerging Internet multimedia technologies. It differentiates itself from other messaging services on the following aspects:

- *Multimedia capabilities*: MMS integrates multimedia features, allowing message contents to be choreographed on the screen of the receiving device. MMS phones typically allow the composition of messages in the form of slideshow presentations composed of sounds, pictures, text and video clips.
- *Electronic mail and phone number addressing modes*: MMS supports several addressing models, including the Internet addressing mode (e.g. gwenael@lebodic.net for an Internet user) and the phone number addressing mode (e.g. +33607080402 for a mobile user). Consequently, a message can be addressed to a recipient using an email address or a phone number.
- *Efficient transport mechanisms*: MMS relies on an efficient message retrieval mechanism. When a message is awaiting retrieval, it is stored temporarily on the network side. The network provides a short notification to the recipient mobile device, indicating that a message awaits retrieval. The mobile device can then automatically fetch the message and notify the user of the reception of a new message. Alternatively, the mobile device can notify the user that a message is awaiting retrieval, and it becomes the user's responsibility to retrieve the message manually at his/her own convenience. Up to now, communications between the MMS phone and the network are performed with binary protocol data units instead of text-based transactions as commonly found over the Internet. This leads to a more optimal use of scarce radio resources.
- *Charging framework*: Charging is of key importance for operators since it allows the generation of users' bills according to the billing model in place. MMS offers an extensive charging framework, which can feed any operator billing system. The charging framework leaves freedom to operators for the development of billing models tailored to market specificities.
- *Future-proof open standards and worldwide acceptance*: Last but not the least, MMS is the result of a collaborative work led by major market players from the mobile industry. MMS technical specifications are developed in open standardization forums

with the continuous objective of designing a future-proof messaging service meeting the requirements of worldwide markets.

## 1.5 Billing Models

The previous sections showed that billing models for Japanese photo messaging services are based on the volume of data required for uploading and retrieving messages. As for Japanese services, the most efficient transport technology for MMS is the packet-based transmission. Nevertheless, billing models for MMS differ from those in place for Japanese messaging services. For MMS, the following billing models are emerging:

- *Flat rate per message sending*: With this billing model, the message sender pays for the cost of sending a message to one or more recipients. The message is free of charge for the receiver. The sender pays a flat rate per message and per recipient (around €0.40 per message for each recipient, regardless of message size[3]). Operators usually offer post-paid charging (e.g. monthly invoice) but also allow pre-paid charging (e.g. pre-paid cards) for MMS. In addition, the operator may request the user to subscribe to a data service for accessing the messaging service. The situation is more complex for roaming users where the roaming sender is typically charged a higher fee for sending a message (around €1 per message for each recipient) and a roaming user may also be charged for receiving a message (around €1 per retrieved message while roaming).
- *Variable rate based on message content class per message sending*: This billing is similar to the previous one except that the message sender pays according to the message contents, message size and number of message recipients. A set of message content classes have been defined in the MMS standard (text, image basic, image rich, video basic and video rich) and may soon become the basis of such a variable rate billing model. For instance, the user may be charged €0.40 for a message containing an image only (up to 30 kB) and €1 for a message containing a video clip (up to 100 kB).
- *Subscription*: With this billing model, the user pays a monthly fee and does not pay for sending or retrieving messages. The operator may limit the number of messages that can be sent for a given period of time.

For communications between mobile subscribers, the most prominent billing model for MMS was initially the one with a flat rate per message sending. Operators favour this model that has proved its efficiency for SMS. However, message sizes are growing and message contents are becoming more sophisticated (e.g. video clips), and operators are now adopting a billing model based on message content classes. The billing of value-added services over MMS is usually based on service subscriptions

---

[3] Usually with a maximum message size of 30 kB for initial MMS implementations.

(e.g. monthly subscription fee), unless the service is subsidized by advertisement. In the latter case, the service becomes free of charge for the end-user.

Note that MMS interworking between operators is far from being guaranteed, and sending a multimedia message from one MMS domain to another is sometimes not allowed. If the message recipient belongs to another MMS domain, then it is common for the operator to send a short text message (e.g. SMS) to the recipient instead. The short message contains a URL pointing to a server from which a WML/XHTML page representing the multimedia message can be fetched with any WAP browser.

## 1.6 Usage Scenarios

Two types of usage scenarios are initially targeted for MMS: person-to-person messaging and content-to-person messaging. The person-to-person scenario is the prominent use case for the first wave MMS. The second wave MMS should see the emergence of new innovative content-to-person services.

### 1.6.1 Person-to-person Messaging

The use of the Multimedia Messaging Service in the person-to-person scenario is tightly associated with the availability of multimedia accessories such as a digital camera or a camcorder. These multimedia accessories may be built into the mobile handset as shown in Figure 1.1 or provided as external accessories that can be connected to the phone. They are used to capture still images and video clips to be inserted in multimedia messages. In this category, photo messaging refers to the typical scenario where the subscriber takes a snapshot of a scene while on the move and sends it as part of a multimedia message to one or more recipients.

The user usually has the possibility to send the message to one or more recipients belonging to the following groups:

- *MMS users*: Users who have an MMS phone and the corresponding service subscription.

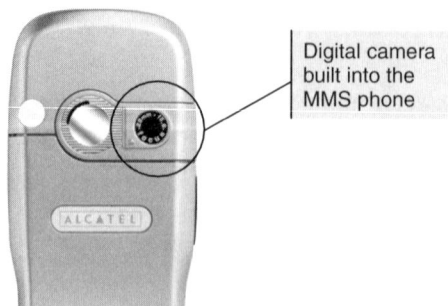

**Figure 1.1**    Built-in camera in MMS phone

- *Users of legacy handsets*: Users who have a legacy phone without support for MMS. For instance, if a user sends a multimedia message (via MMS) to a legacy user, the network can generate a short message and deliver it (via SMS) to the legacy user. The short message contains the address of an Internet page that can be viewed by the legacy user using any embedded WAP browser.
- *Internet users*: Internet users can receive multimedia messages originating from MMS users. Multimedia messages, as generated by MMS phones, are not 'yet' directly understandable by email clients such as Microsoft Outlook or Lotus Notes. To cope with this issue, the multimedia message is transcoded in the MMS domain to a more suitable form understandable by email clients. Note that a transcoded multimedia message may not represent exactly the contents of the original multimedia message. The slideshow structure of multimedia messages is often lost in the transcoding operation. Owing to the low market penetration of MMS-enabled phones, the usage scenario where a mobile user sends a message to an Internet user is currently the most prominent one in the person-to-person domain. However, this is expected to change in the near future with the rising penetration of MMS-enabled phones.

### 1.6.2 Content-to-person Messaging

In the context of MMS, a Value-Added Service (VAS) provider is an organization that offers an added-value service based on MMS. A VAS application may provide weather notifications, news updates, entertainment services, location-based information and so on delivered to the phone as a multimedia message. For this purpose, the provider sets up a VAS application, which generates multimedia messages and sends them to one or multiple recipients via the MMS provider infrastructure. In many cases, the user needs to subscribe first to the value-added service in order to receive corresponding messages. The service can be activated by sending a message to the VAS application. Mass distribution of information can be achieved with a value-added service. In order to operate a value-added service, the VAS provider has to establish a service agreement with the MMS provider. In particular, such an agreement specifies how the revenue generated by the value-added service is shared between the MMS provider and the VAS provider. The content-to-person scenario is also referred to as the machine-to-person scenario.

In the content-to-person scenario, one can distinguish several successful services in the first wave MMS, including the one described below:

- *Delivery of information with Buongiorno*: Buongiorno started with the publication of information using the text-based SMS. Buongiorno has now extended its basket of services with the publication of information using MMS. From the Buongiorno website (http://www.buongiorno.com), a greeting service allows users to pick up an image from a selection, provide a personal text greeting and select an audio clip for the composition of a multimedia message to be delivered to one or more recipients with MMS. At the time of writing, Buongiorno provided services in the following countries: France, Germany, Austria, Italy, Portugal, Spain and the United Kingdom. The primary revenue sources for Buongiorno are user subscription and advertising.

Services in the content-to-person scenario are expected to be further elaborated for the second wave MMS with the interfacing of MMS centres to the Internet world and the possibility to generate appealing contents (e.g. larger messages with video clips).

### 1.6.3 Further Applications

Recently, Kodak announced the availability of an online service allowing mobile users to upload, store, share and order prints of pictures using their MMS phones. With this service, the mobile user takes a photo with the phone-embedded camera and stores it in an online personal photo album. Later, the user can access and retrieve the photo from the album, forwarding it again to other MMS phones, or order prints of the photo.

MMS can be considered as a building block enabling the development of other services. For instance, it can be envisaged to develop embedded monitoring applications that regularly take photos of critical sites and send messages with these photos to a remote monitoring centre. These applications typically address the requirements of niche markets.

## Further Reading

G. Le Bodic, *Mobile Messaging, Technologies and Services: SMS, EMS and MMS*, John Wiley & Sons, Chichester, 2002.

T. Natsuno, *i-mode Strategy*, John Wiley & Sons, Chichester, 2002.

# 2

# Standardization of MMS

Standardization of telecommunications technologies and associated service enablers is of key importance for the development of devices in a multi-vendor environment. In the context of Multimedia Messaging Services (MMS), standards allow mobile devices and network elements to interoperate in an efficient manner. Standardization of MMS does not mean designing from scratch all technologies required for enabling interoperable communications. Instead, standardization means identifying the most appropriate elements in the basket of existing technologies in order to allow a rapid roll-out of the service, creating new technologies only when no appropriate solution exists.

Compared with other mobile messaging services such as the Short Message Service (SMS) and the Enhanced Messaging Service (EMS), the standardization picture for MMS has become very complex. Several standardization organizations have collaborated in order to produce stable technical specifications for MMS in a timely manner. Organizations that have been actively involved in the design of MMS standards are the Third Generation Partnership Project (3GPP) and the Wireless Application Protocol (WAP) Forum. Since 2002, the WAP Forum has merged with other bodies to form the Open Mobile Alliance (OMA). Consequently, MMS activities of the WAP Forum have now been transferred to OMA. Most MMS standards produced by these organizations partially rely on existing technologies developed by bodies such as World Wide Web Consortium (W3C) and Internet Engineering Task Force (IETF).

For any engineer involved in designing solutions based on MMS standards, it becomes essential to acquire a basic understanding on how standardization bodies proceed to produce standards. Most importantly, engineers need to identify dependencies linking MMS standards among themselves and understand how standards get created, reach a mature stage and evolve over time. For this purpose, this chapter introduces the working procedures of organizations outlined below and provides an insight of their organizational structure in terms of working groups. Rules for numbering/referencing MMS standards are explained and illustrated with examples.

- *Third Generation Partnership Project (3GPP)*: 3GPP is not a standardization development organization in its own right but rather a joint project between several

---

*Multimedia Messaging Service: An Engineering Approach to MMS*   Gwenaël Le Bodic
© 2003 John Wiley & Sons, Ltd   ISBN: 0-470-86253-X

regional standardization bodies from Europe, North America, Korea, Japan and China. The prime objective of 3GPP is to develop UMTS technical specifications. It is also responsible for maintaining existing GSM specifications and developing further GSM extensions (e.g. GPRS). This encompasses the development of widely accepted technologies and service capabilities. 3GPP is strongly involved in the development of MMS standards (general service requirements, architecture, formats and codecs and several low-level technical realizations).

- *Third Generation Partnership Project 2 (3GPP2)*: 3GPP2 is another standardization partnership project established out of the International Telecommunication Union's (ITU) International Mobile Telecommunications "IMT-2000" initiative. The role of 3GPP2 is to produce specifications for industrial players from North American and Asian markets with focus on next-generation CDMA networks. In the scope of this project, 3GPP2 looks at refining requirements for MMS and designing alternative realizations of interfaces defined in 3GPP and OMA standards.

- *WAP Forum*: The Wireless Application Protocol (WAP) Forum was a joint project for the definition of WAP technical specifications. This encompassed the definition of a framework for the development of applications to be executed in various wireless platforms. The WAP Forum produced the initial MMS standards for the support of MMS in the WAP environment.

- *Internet Engineering Task Force (IETF)*: IETF is a large community of academic and industrial contributors that defines the protocols in use on the Internet.

- *World Wide Web Consortium (W3C)*: W3C is a standardization body that concentrates on the development of protocols and formats to be used in the World Wide Web. Well-known formats and protocols published by W3C are the Hypertext Transfer Protocol (HTTP) and the eXtensible Modelling Language (XML).

- *Open Mobile Alliance (OMA)*: OMA is a standardization forum established in June 2002. Activities of several existing standardization bodies including the ones of the WAP Forum (MMS and others) have been transferred to OMA. OMA is therefore actively involved in maintaining MMS standards designed by the WAP Forum and producing new standards for next generations of MMS devices.

## 2.1 MMS Standards

MMS is a sophisticated multimedia messaging service and has required a tremendous standardization workload. Several standardization bodies have therefore collaborated in order to produce the technical specifications to allow the introduction of MMS-capable devices on the market at the most appropriate time. In this configuration, 3GPP has taken the lead in identifying the high-level service requirements, designing the MMS architecture, producing several low-level technical realizations and identifying appropriate codecs/formats and streaming protocols. On the other hand, the WAP Forum took the responsibility of defining the low-level technical realizations of the interface bridging the MMS phone and the network in the WAP environment. Additionally, a group of telecommunications vendors, known as the MMS-IOP group, also produced specifications (the MMS conformance document) to guarantee the interoperability between first MMS devices. In 2002, MMS activities of the WAP Forum and the MMS-IOP group were merged into OMA to allow a more efficient standardization development process for MMS. Of course, 3GPP and the WAP Forum/OMA did not produce all MMS specifications from scratch and did manage to build up MMS standards on the basis of existing proven standards such as the ones produced by W3C and IETF, developing new technologies only when not available elsewhere.

Regarding this collaborative work, the standardization picture for MMS is becoming more and more complex as the MMS standards evolve. It becomes difficult for developers of applications to understand the dependencies linking MMS standards produced by different organizations and to value the level of maturity of available standards.

## 2.2 Third Generation Partnership Project

The European Telecommunications Standard Institute (ETSI) and the Conférence Européenne des Postes et Télécommunications (CEPT) have carried out work on the GSM standards during a period of almost 18 years. Within the scope of the ETSI standardization organization, the work was carried out by the Special Mobile Group (SMG) technical committee. In 2000, the committee agreed to transfer the responsibility of the development and maintenance of the GSM standards to the Third Generation Partnership Project. 3GPP was set up in 1998 by five standard development organizations (including ETSI) with the objective of collaborating on the development of interoperable mobile systems (a sixth organization joined the partnership later). The six organizations represent telecommunications companies from five different parts of the world:

- European Telecommunications Standards Institute (ETSI) for Europe.
- Committee T1 for the United States.
- Association of Radio Industries and Businesses (ARIB) for Japan.
- Telecommunications Technology Committee (TTC) for Japan.
- Telecommunications Technology Association (TTA) for Korea.
- China Wireless Telecommunication Standard (CWTS) for China.

Each individual member of one of the six partners can contribute to the development of 3GPP specifications. In order to define timely services and technologies, individual members are helped by several market representative partners. At the time of writing this book, 3GPP market representatives were the UMTS Forum,[1] the Global mobile Suppliers Association (GSA),[2] the GSM Association (GSMA),[3] the IPv6 Forum,[4]

---

[1] http://www.umts-forum.org/
[2] http://www.gsacom.com/
[3] http://www.gsmworld.com/
[4] http://www.ipv6forum.com/

the 3G.IP focus group[5] and the 3G Americas.[6] The responsibility of these market representative partners consists of identifying requirements for 3G services. In this process, the six partner organizations take the role of publishers of 3GPP specifications. It has to be noted that large parts of the 3GPP work, such as SMS and MMS, are also applicable to 2G and 2.5G systems.

### 2.2.1 3GPP Structure

The 3GPP standardization process strictly defines how partners should coordinate the standardization work and how individual members should participate in the development of specifications. There is a clear separation between the coordination work of 3GPP partners and the development of specifications by individual members. This separation enables a very efficient and robust standardization process. In order to achieve it, the 3GPP structure is split into the Project Coordination Group (PCG) and five Technical Specifications Groups (TSGs). The PCG is responsible for managing and supervizing the overall work carried out within the scope of 3GPP, whereas TSGs create and maintain 3GPP specifications. PCG and TSGs endeavour to reach consensus on all issues. However, decisions in PCG and TSGs can be made by vote if consensus cannot be reached. In each TSG, several working groups (WGs) create and manage specifications for a set of related technical topics (for instance CN WG5 deals with the set of technical topics related to the Open Service Architecture). If the set of technical topics is too broad, then a WG may be further split into Sub Working Groups (SWGs). This is the case for T WG2 (or also T2 for short), which deals with mobile terminal services and capabilities. T2 is split into three SWGs:

- T2 SWG1 deals with the Mobile Execution Environment (MExE).
- T2 SWG2 deals with user equipment capabilities and interfaces.
- T2 SWG3 deals with messaging aspects. Activities of sub-working group T2 SWG3 encompass the development of messaging services and technologies including SMS, EMS, Cell Broadcast Service and MMS.

Figure 2.1 shows the list of 3GPP TSGs and corresponding WGs. Note that all TSGs are responsible for their own work items and specifications. However, TSG SA, being responsible for the overall architecture and service capabilities, has an additional responsibility for cross TSG coordination.

### 2.2.2 3GPP Specifications: Release, Phase and Stage

Documents produced by the 3GPP are known as specifications. Specifications are either Technical Specifications (TS) or Technical Reports (TR). Technical specifications define a GSM/UMTS standard and are published independently by the six

---

[5] http://www.3gip.org/
[6] http://www.3gamericas.org/

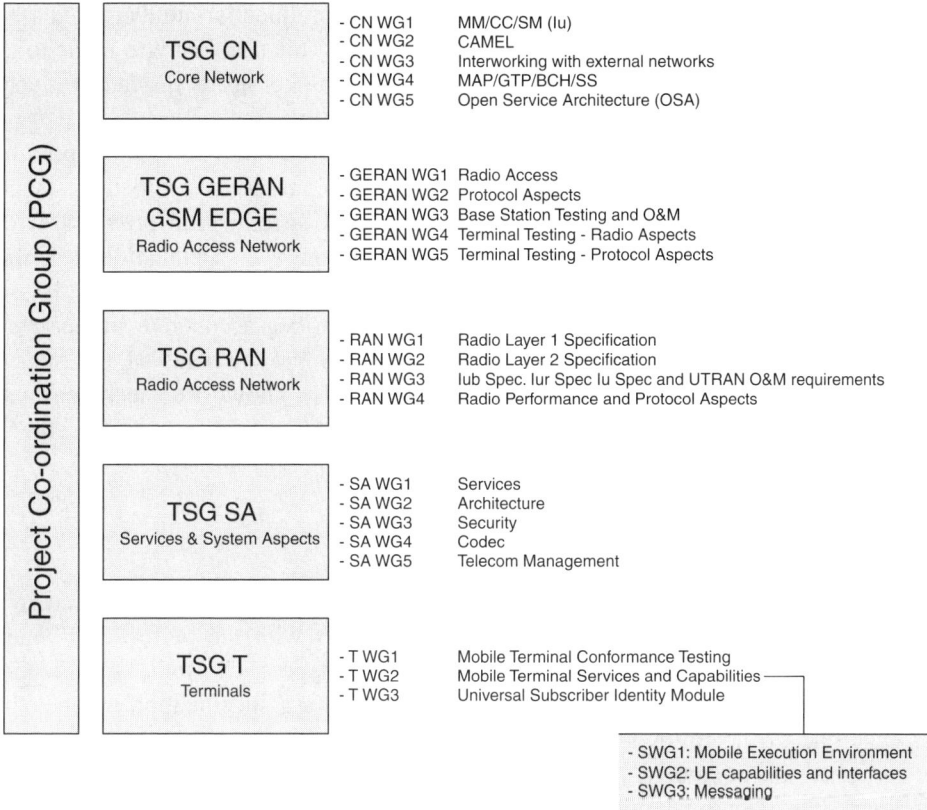

**Figure 2.1** 3GPP structure

partners (ETSI, Committee T1, ARIB, TTC, TTA and CWTS). Technical reports are, for example, feasibility studies for new features/services and they sometimes become technical specifications later. In order to fulfil ever-changing market requirements, 3GPP specifications are regularly extended with new features. To ensure that market players have access to a stable platform for implementation and, meanwhile, to allow the addition of new features, the development of 3GPP specifications is based on a concept of parallel releases. In this process, specifications are regularly frozen. Only essential corrections are permitted for a frozen specification. New work can still be carried out but will be incorporated in the next release of the same specification. An engineer implementing a commercial solution based on one or more 3GPP standards should, as much as possible, base the work on frozen specifications. An unfrozen specification is subject to change and should never be considered as a stable platform on which to build a commercial solution. In 3GPP, technical specifications are typically frozen in intervals of one to one-and-a-half year. Consequently, releases used to be named according to the expected specification freezing date (Release 98, Release 99, etc.). In 1999, the 3GPP decided that releases produced after 1999 would no longer be named according to the year but according to a unique sequential number (Release 5 followed Release 4, which itself followed Release 99). This decision was made to get more flexibility in adjusting the timing of releases to market needs instead of always having one release per year.

Each 3GPP technical specification is usually categorized into one of three possible stages. The concept of characterizing telecommunication services into three stages was first introduced by ITU in [ITU-I.130]. A stage 1 specification provides a service description from a service-user's perspective. A stage 2 specification describes a logical analysis of the problem to be solved, a functional architecture and associated information flows. A stage 3 specification describes a concrete implementation of the protocols between physical elements onto the elements of the stage 2 functional architecture. A stage 3 implementation is also known as a technical realization. Note that several technical realizations may derive from a common stage 2 specification.

### 2.2.3 3GPP Specifications: Numbering Scheme

Each 3GPP technical document (report or specification) is uniquely identified by a reference as shown in Figure 2.2. The reference starts with the prefix '3GPP' and is followed by two letters identifying the document type ('TS' for a specification and 'TR' for a report). After the document type, follows a specification number that can take one of the following forms: *aa.bbb* or *aa.bb*. In the specification number, *aa* indicates the document's intended use as shown in Table 2.1. In the document reference, the document number is followed by a version number in the format *Vx.y.z*. In this format, $x$ represents the release, $y$ represents the technical version and $z$ represents the editorial version. Table 2.2 shows how the document version is formatted according to its associated release. The freezing date for each release is also indicated.

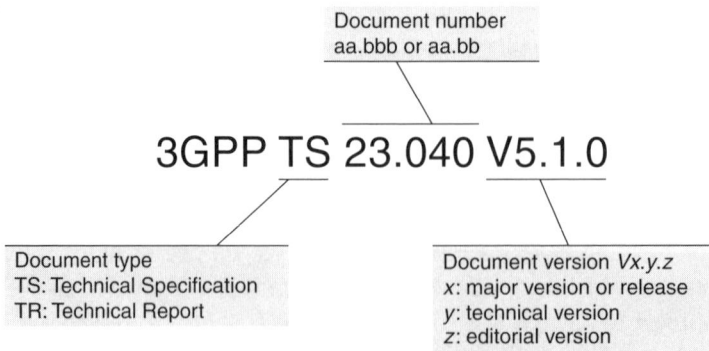

Document number
aa.bbb or aa.bb

## 3GPP TS 23.040 V5.1.0

Document type
TS: Technical Specification
TR: Technical Report

Document version *Vx.y.z*
*x*: major version or release
*y*: technical version
*z*: editorial version

**Figure 2.2**   3GPP specification type, number and version

**Table 2.1**   3GPP specifications/numbering scheme

| Range for GSM up to and including Release 99 | Range for GSM Release 4 onwards | Range for UMTS Release 99 onwards | Type of use |
|---|---|---|---|
| 01.bb | 41.bbb | 21.bbb | Requirement specifications |
| 02.bb | 42.bbb | 22.bbb | Service aspects |
| 03.bb | 43.bbb | 23.bbb | Technical realizations |
| 04.bb | 44.bbb | 24.bbb | Signalling protocols |
| 05.bb | 45.bbb | 25.bbb | Radio access aspects |
| 06.bb | 46.bbb | 26.bbb | Codecs |
| 07.bb | 47.bbb | 27.bbb | Data |
| 08.bb | 48.bbb | 28.bbb | Signalling protocols |
| 09.bb | 49.bbb | 29.bbb | Core network signalling protocols |
| 10.bb | 50.bbb | 30.bbb | Programme management |
| 11.bb | 51.bbb | 31.bbb | SIM/USIM |
| 12.bb | 52.bbb | 32.bbb | Charging and OAM&P |
| 13.bb | | | Regulatory test specifications |
| | | 33.bbb | Security aspects |
| | | 34.bbb | Test specifications |
| | | 35.bbb | Algorithms |

**Table 2.2**   3GPP specifications/releases

| GSM/edge release | 3G release | Abbreviated name | Spec. number format | Spec. version format | Freeze date |
|---|---|---|---|---|---|
| Phase 2 + Release 6 | Release 6 | Rel-6 | aaa.bb (3G) | 6.x.y (3G) | March 2004 |
| Phase 2 + Release 5 | Release 5 | Rel-5 | aa.bb (GSM) | 5.x.y (GSM) | March 2002 |
| Phase 2 + Release 4 | Release 4 | Rel-4 | aaa.bb (3G) aa.bb (GSM) | 4.x.y (3G) 9.x.y (GSM) | March 2001 |
| Phase 2 + Release 99 | Release 99 | R99 | aaa.bb (3G) aa.bb (GSM) | 3.x.y (3G) 8.x.y (GSM) | March 2000 |
| Phase 2 + Release 98 | | R98 | aa.bb | 7.x.y | Early 1999 |
| Phase 2 + Release 97 | | R97 | aa.bb | 6.x.y | Early 1998 |
| Phase 2 + Release 96 | | R96 | aa.bb | 5.x.y | Early 1997 |
| Phase 2 | | PH2 | aa.bb | 4.x.y | 1995 |
| Phase 1 | | PH1 | aa.bb | 3.x.y | 1992 |

In addition to its reference, a title is also provided for a 3GPP document. For instance, the following document contains the definition of MMS stage 1:

Reference: 3GPP TS 22.140 V5.2.0

Title: Multimedia Messaging Service, Stage 1

Lists of available 3GPP specifications are provided in the documents listed in Table 2.3.

**Table 2.3** Specifications listing the GSM/UMTS specifications produced by 3GPP

| Release | List of GSM specifications | List of UMTS specifications |
|---|---|---|
| Release 99 | [3GPP-01.01] | [3GPP-21.101] |
| Release 4 | [3GPP-41.102] | [3GPP-21.102] |
| Release 5 | [3GPP-41.103] | [3GPP-21.103] |

3GPP specifications can be downloaded from the 3GPP website at http://www.3gpp.org.

## 2.3 Third Generation Partnership Project 2

The Third Generation Partnership Project 2 (3GPP2) is another standardization partnership project established out of the International Telecommunication Union's (ITU) International Mobile Telecommunications "IMT-2000" initiative. The role of 3GPP2 is to produce specifications for industrial players from North American and Asian markets with focus on next-generation CDMA networks. In the scope of this project, 3GPP2 looks at refining requirements for MMS and designing alternative realizations of interfaces defined in 3GPP and OMA standards.

3GPP2 specifications can be downloaded from the 3GPP2 website at http://www.3gpp2.org.

## 2.4 WAP Forum Specifications

Prior to its integration in OMA, the WAP Forum concentrated on the definition of a generic platform for the development of applications for various wireless technologies. The WAP Forum was organized into functional areas as shown in Figure 2.3.

The WAP Forum used to manage four types of technical documents:

- *Specification*: A specification contains technical or procedural information. At any given time, a specification is associated with a stage such as proposal, draft and so on. This stage indicates the level of maturity of the specification content.
- *Change Request (CR)*: An unofficial proposal to change a specification. A change request is proposed by one or more individuals for discussion between WAP Forum members.
- *Specification Change Document (SCD)*: An SCD is the draft of a proposed modification of a specification. An SCD can only be produced by the specification working group responsible for the corresponding specification. An SCD applies to a specific version of a specification.
- *Specification Implementation Note (SIN)*: An SIN is an approved modification of a previously published specification. SINs are used to fix bugs or to revise an existing approved specification. A SIN applies to a specific version of a specification.

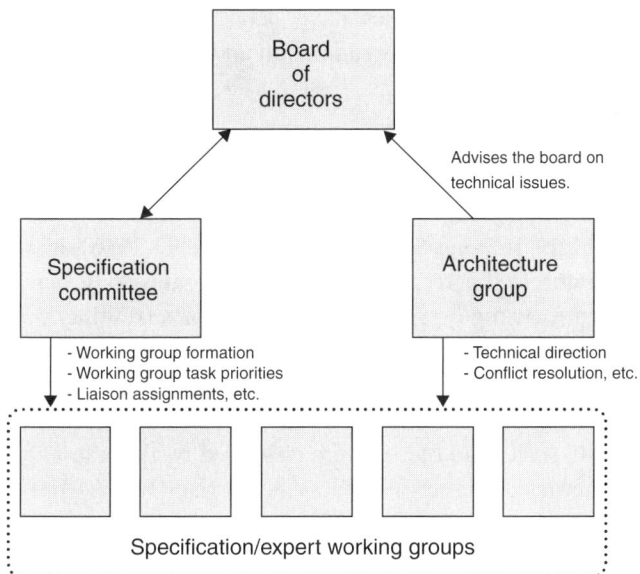

**Figure 2.3** WAP forum organization

A WAP Forum document is identified by a Document IDentifier (DID). A specification keeps its associated DID for its entire lifespan (all revisions of the specification and the approved specification).

WAP Forum specifications are named according to the convention outlined in Figure 2.4.

Only approved specifications should be considered as a basis for the development of WAP-based solutions. OMA members can download WAP Forum specifications from the OMA website at http://www.openmobilealliance.org.

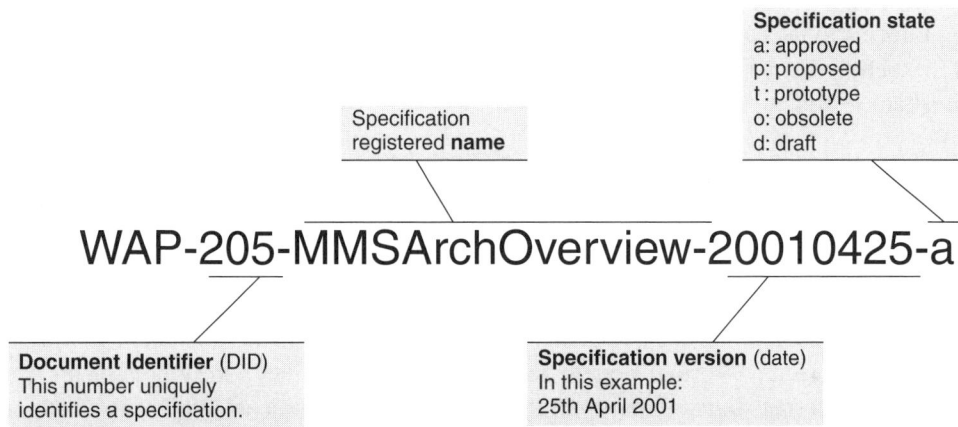

**Figure 2.4** WAP Forum specification numbering

In 2002, all WAP Forum activities have been transferred to the Open Mobile Alliance. Section 2.7 introduces the organization and working procedures of the Open Mobile Alliance.

## 2.5 Internet Engineering Task Force

The IETF produces technical specifications related to Internet protocols and procedures in the scope of the Internet Standards Process [RFC-2026]. It is an international forum open to any interested party. In IETF, the technical work is carried out by working groups, each one focusing on specific technical topics (routing, transport, security, etc.). Furthermore, working groups are categorized into areas managed by area directors. Area directors are members of the Internet Engineering Steering Group (IESG). In IETF, the Internet Architecture Board (IAB) provides an architectural oversight of the Internet. Both the IESG and the IAB are chartered by the Internet Society (ISOC). In addition, the Internet Assigned Numbers Authority (IANA) has the responsibility for assigning unique numbers for Internet protocols (applications ports, content types, etc.).

### 2.5.1 IETF Documents

IETF communications are primarily performed through the publication of a series of documents known as Request For Comments (RFCs). First RFCs on Internet networking were produced in 1969 as part of the original ARPA wide-area networking (ARPANET) project. RFCs are numbered in chronological order of creation. An RFC, documenting an Internet standard, is given an additional reference in the form STD*xxx* and becomes part of the STD subseries of the RFC series. RFCs are classified into five high-level categories:

1. Standard track (including proposed standards, draft standards and Internet standards)
2. Best current practices
3. Informational RFCs
4. Experimental RFCs
5. Historic RFCs.

For instance, the following specification is published as RFC822. The specification is an Internet standard also known as STD0011:

```
RFC822  Standard for the format of ARPA Internet text messages.
     D.Crocker. Aug-13-1982 / Status: STANDARD / STD0011
```

During the development of a specification, draft versions of the document are often made available for informal review and comments. These temporary documents are known as Internet drafts. Internet drafts are working documents subject to changes without notice. Consequently, Internet drafts should not be considered as stable specifications on which to base developments.

### 2.5.2 Internet Standard Track

Specifications subject to the Internet Standards Process fall into two categories: technical specifications and applicability statements. A technical specification is a description of a protocol, service, procedure, convention or format. On the other hand, an applicability statement indicates how one or more technical specifications may be applied to support a particular Internet capability. Specifications that are to become Internet standards evolve over a scale of maturity levels known as the standard track. The standard track is composed of three maturity levels:

- *Proposed standard*: This is the specification at the entry level. The IESG is responsible for registering an existing specification as a proposed standard. A proposed standard is a specification that has reached a stable stage and has already been the subject of public review. However, implementers should consider proposed standards as immature specifications.
- *Draft standard*: A proposed standard can be moved to the draft standard of the standard track if at least two implementations based on the specification exist and interoperate well. Since draft standards are considered as almost final specifications, implementers can safely design services on the basis of draft standards.
- *Internet standard*: An Internet standard (also referred to as a standard) is a specification that is the basis of many significant implementations. An Internet standard has reached a high level of technical maturity.

RFCs can be downloaded from the IETF website at http://www.ietf.org.

## 2.6 World Wide Web Consortium

The World Wide Web Consortium is a standardization body, created in 1994, involved in the development of widely accepted protocols and formats for the World Wide Web. Technical specifications published by W3C are known as recommendations. W3C collaborates closely with IETF. W3C activities are organized into groups: working groups (for technical developments), interest groups (for more general work) and coordination groups (for communications among related groups). W3C groups are organized into five domains:

- The *Architecture domain* includes activities related to the development of technologies that represent the basis of the World Wide Web architecture.
- The *Document formats domain* covers all activities related to the definition of formats and languages.
- The *Interaction domain* includes activities related to the improvements of user interactions with the World Wide Web. This includes the authoring of content for the World Wide Web.
- The *Technology and society domain* covers activities related to the resolution of social and legal issues along with handling public policy concerns.

- The *Web accessibility initiative* aims at promoting a high degree of usability for disabled people. Work is carried out in five primary areas: technology, guidelines, tools, education and outreach, and research and development.

To date, significant W3C contributions include the architecture of the initial World Wide Web (based on HTML, URIs and HTTP), XML, XHTML, SVG and SMIL. W3C follows a dedicated recommendation track to initiate discussions on proposed technologies and to ultimately publish recommendations. The recommendation track defines four technical specification statuses corresponding to four increasing levels of maturity. These statuses are depicted in Figure 2.5.

A *working draft* is the initial status for a technical specification in the W3C recommendation track. It is a work item that is being or will be discussed within the relevant W3C working group. However, the status 'working draft' for a technical specification does not imply that there is a consensus between W3C members on the acceptability of the proposed technology.

A *last call working draft* is a special working draft that is regarded by the relevant working group as fulfilling the requirements of its charter. It has the status 'last call working draft' when the relevant working group seeks technical review from other W3C groups, W3C members and the public.

A *candidate recommendation* is a technical specification that is believed to fulfil the relevant working group's charter, and has been published in order to gather implementation experience and feedback.

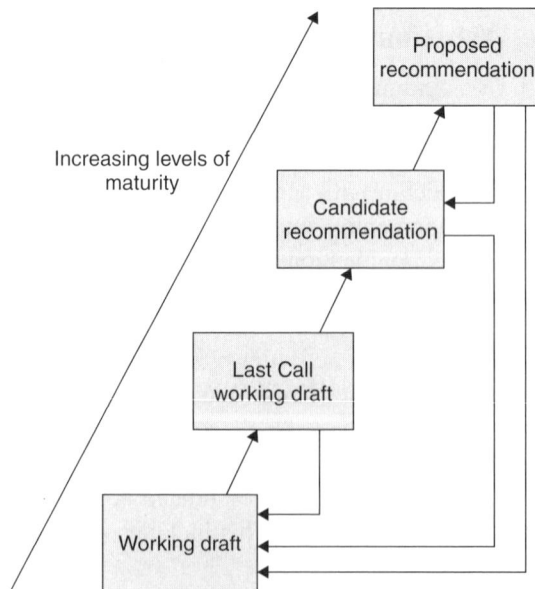

**Figure 2.5**   W3C recommendation track

Finally, a *proposed recommendation* is a technical specification that has reached the highest level of maturity in the W3C recommendation track. It obviously fulfils the requirements of the relevant working group's charter but has also benefited from sufficient implementation experience and has been carefully reviewed. Only a proposed W3C recommendation can be considered as a stable technical specification on which to base the development of commercial solutions.

W3C technical specifications can be retrieved from http://www.w3c.org.

## 2.7 Open Mobile Alliance

The Open Mobile Alliance (OMA) is a new standardization forum established in June 2002 by nearly 200 companies representing the whole mobile services value chain. It is chartered to develop interoperable application enablers for the mobile industry. It designs specifications for applications enablers, which are bearer-agnostic and independent from any operating system. OMA was not created from scratch but was rather organized as a merge of several existing standardization forums. These forums, with sometimes overlapping activities, included the WAP Forum, the Wireless Village (WV), the MMS Interoperability Group (MMS-IOP), the SyncML Initiative, the Location Interoperability Forum (LIF), the Mobile Wireless Internet Forum (MWIF) and the Mobile Games Interoperability Forum (MGIF).

### 2.7.1 OMA Organization

In the OMA organization, the technical plenary is a chartered standing committee of the board of directors. The technical plenary is responsible for technical specification drafting activities, approval and maintenance of technical specifications and resolution of technical issues. The organization of OMA is depicted in Figure 2.6.

The task of OMA working groups (WG) is to accomplish the technical work as defined by the technical plenary. On the other hand, committees are responsible for establishing rules for OMA operations and processes and for controlling the release of OMA specifications. Responsibilities of OMA working groups and committee are described below:

- *Requirements (REQ)*: This WG is responsible for identifying the use cases for services and identifying interoperability and usability requirements.
- *Architecture (ARCH)*: This WG is in charge of the design of the overall OMA system architecture.
- *Mobile Application Group (MAG)*: This WG is responsible for building application enablers including browsing, synchronization, IMPS, location and MMS. Activities of this WG are delegated to sub-working groups. The MMS Sub-working Group (MMSG) is responsible for the design of OMA MMS standards.
- *Mobile Web Services (MWS)*: This WG is responsible for defining application enablers for web services in OMA.

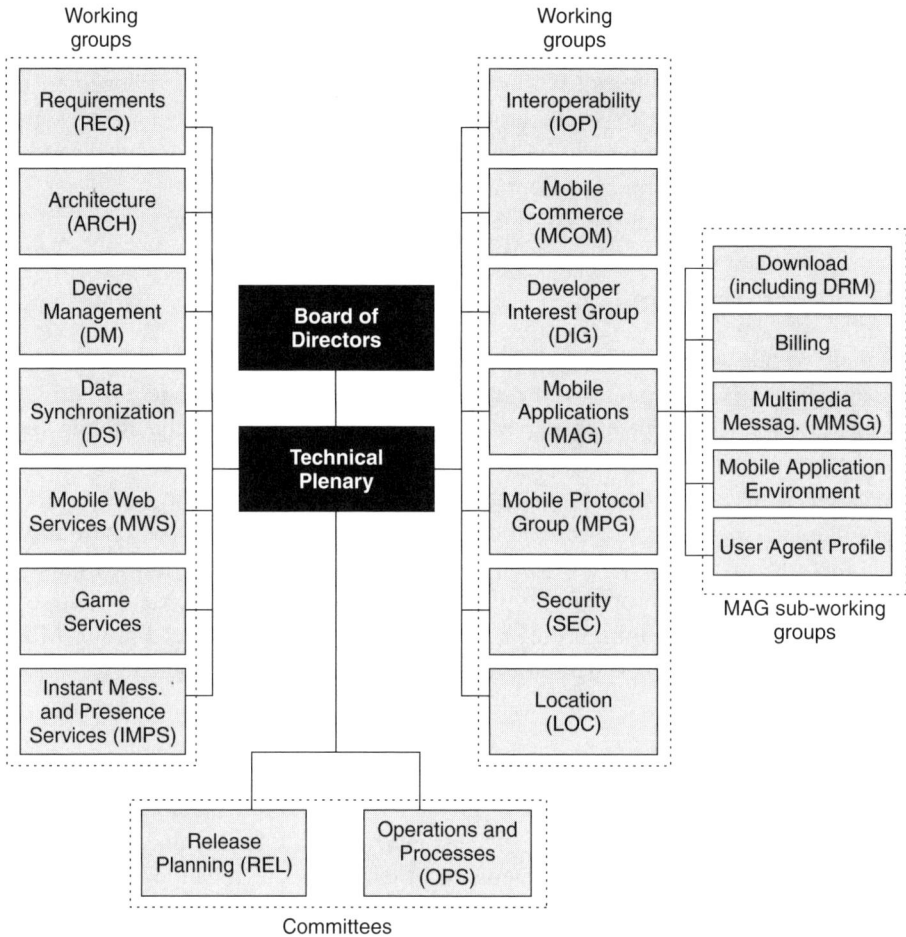

**Figure 2.6**  OMA organization

- *Instant Messaging and Presence Services (IMPS)*: The IMPS WG's objectives are to identify, specify and maintain the requirements, architecture, technical protocol/format/interface and interoperability specifications for mobile instant messaging and presence services.
- *Interoperability (IOP)*: This WG focuses on testing interoperability and solving identified issues. Activities of this WG are delegated to sub-working groups (browsing, synchronization, IMPS, location, MMS, etc.). It also organizes test events as further described in Chapter 7.
- *Mobile Protocols Group (MPG)*: This WG is responsible for providing a consolidated and consistent approach within OMA towards the use of generic bearers and protocols (including WSP, WTP and DNS).

- *Device Management (DM)*: This WG is in charge of specifying protocols and mechanisms that achieve management of devices (e.g. configuration settings, operating parameters, software installation and parameters, application settings, user preferences).
- *Data Synchronization (DS)*: This WG designs specifications for data synchronization (including but not limited to the SyncML technology).
- *Game services*: This WG is responsible for defining interoperability specifications, application programming interfaces (APIs) and protocols for network-enabled gaming.
- *Mobile Commerce (MCOM)*: This WG aims at producing standards for technologies enabling mobile commerce in order to meet the demands of banking and financial industries, retailers and mobile users.
- *Security (SEC)*: This WG focuses on specifying the operation of adequate security mechanisms, features and services by mobile clients, server and related entities.
- *Location (LOC)*: This WG designs an end-to-end architectural framework with relevant application and contents interfaces, privacy and security, charging and billing, and roaming for location-based services.
- *Developer Interest Group (DIG)*: In this group, OMA software developers are able to express their requirements into OMA as input for other working groups.

In addition to the working groups, two committees are part of the OMA organization as described below:

- *Operations and Processes (OPS)*. This committee provides support for OMA operation and process activities. This includes the support for liaising with other forums, the assessment for IT requirements and staffing needs and so on.
- *Release Planning (REL)*: This committee is responsible for planning and managing OMA releases according to OMA specifications and interoperability testing (IOT) programmes.

### 2.7.2 OMA Specifications

The OMA process for publishing public technical specifications is based on the delivery of enabler releases and interoperability releases. An enabler release is a set of specifications required for enabling the realization of a service such as MMS, browsing, digital rights management and so on. OMA technical specifications in a draft stage are only made available to OMA members and should not be considered as mature inputs for the design of commercial solutions. When a set of technical specifications has gained enough maturity to be considered for the development of commercial solutions, then OMA publicly releases the set of specifications in the form of an enabler release. Enabler releases evolve over a scale of two maturity phases as shown below:

- *Candidate enabler release (phase 1)*: A candidate enabler release is an approved set of OMA specifications forming the basis of product implementations, which can be tested for interoperability.
- *Approved enabler release (phase 2)*: An approved enabler release has passed phase 1 (candidate release), and associated interoperability test cases have been designed by OMA (see Chapter 7).

Additionally, multiple approved enabler releases (phase 2) can be grouped into a single interoperability release (phase 3). For this purpose, a set of devices conforming to the approved enabler releases are required to pass end-to-end interoperability tests.

An OMA specification is uniquely identified as shown in Figure 2.7.

### 2.7.3 Available Documents

At the time of writing, several sets of OMA specifications had been made publicly available. This includes the following candidate enabler releases (phase 1) – August 2003 status:

- OMA Billing Framework v1.1 (2 documents)
- OMA Browsing v2.1 (18 documents)
- OMA Client Provisioning v1.1 (6 documents)
- OMA DNS v1.0 (2 documents)
- OMA Digital Rights Management v1.0 (4 documents)
- OMA Download v1.0 (2 documents)
- OMA Email Notification v1.0 (2 documents)
- OMA IMPS v1.2 (17 documents)
- OMA Multimedia Messaging Service v1.1 (5 documents)

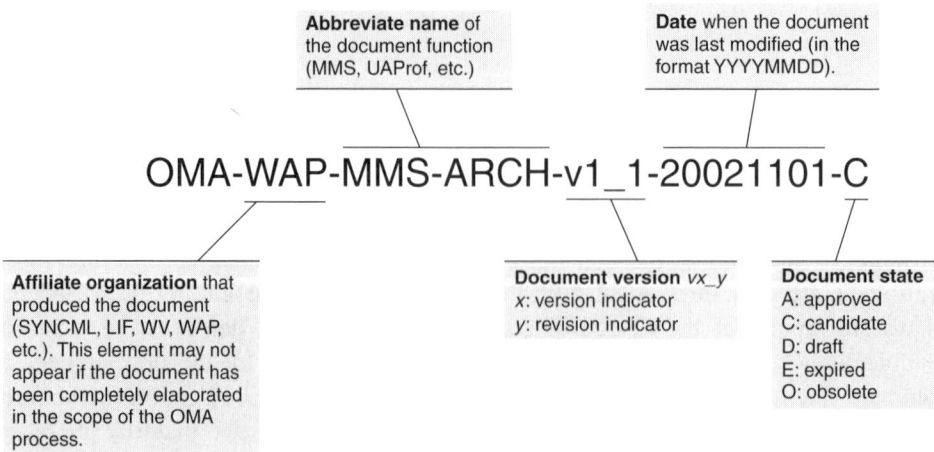

**Abbreviate name** of the document function (MMS, UAProf, etc.)

**Date** when the document was last modified (in the format YYYYMMDD).

## OMA-WAP-MMS-ARCH-v1_1-20021101-C

**Affiliate organization** that produced the document (SYNCML, LIF, WV, WAP, etc.). This element may not appear if the document has been completely elaborated in the scope of the OMA process.

**Document version** $vx\_y$
$x$: version indicator
$y$: revision indicator

**Document state**
A: approved
C: candidate
D: draft
E: expired
O: obsolete

**Figure 2.7**    OMA specification numbering

- OMA User Agent Profile v1.1 (3 documents)
- OMA User Agent Profile v2.0 (2 documents).

OMA IMPS v1.1 (17 documents) is also publicly available as an approved enabler release (phase 2).

OMA technical specifications can be downloaded from http://www.openmobile alliance.org.

## 2.8 Standardization Roadmap for MMS

As shown in this chapter, the roadmap of MMS standards is rather complex. Standards provide different levels of technical information to allow MMS experts to build interoperable implementations, while always improving the existing enabling technologies. Some standards describe the high-level service requirements from which derive other standards dealing with service architecture and interactions between MMS devices. Other standards identify formats and codecs used in the context of MMS, whereas some others concentrate on billing and charging aspects. 3GPP and OMA have designed major MMS standards required for designing MMS solutions. These standards rely on existing generic standards developed by W3C and IETF. Figure 2.8 presents a general organization of 3GPP and OMA MMS standards around the four following specification sets:

- MMS requirement specifications, service aspects and technical realizations.
- MMS codecs and support of streaming.
- MMS charging aspects.
- MMS-related files in the SIM/USIM.

MMS standards become more and more feature-rich as they evolve over time. At appropriate times, standards are frozen at a given level of features in order to allow vendors to produce compliant devices. In the meantime, standardization engineers carry on the work on a new set of additional features for the next generation of MMS devices. At the time of writing, three levels of features were available for the implementation of MMS solutions. These three levels of MMS features are known as MMS 3GPP Release 99, Release 4 and Release 5 corresponding respectively to WAP Forum MMS 1.0, OMA MMS 1.1 and OMA MMS 1.2. Features corresponding to these three levels are summarized in Table 2.4.

Each standard organization proposes a set of several maturity levels over which standards evolve over time from a draft level to a level of higher maturity (frozen for 3GPP and approved for OMA). Figure 2.9 shows the availability of MMS standards according to the corresponding level of features (3GPP release/OMA version).

MMS standards can be downloaded from the websites of respective standardization bodies, as shown in this chapter. Each relevant standard is introduced in Table 2.5.

An online resource page points to all MMS-related standards. This page is available from this book companion website at http://www.lebodic.net/mms_resources.htm.

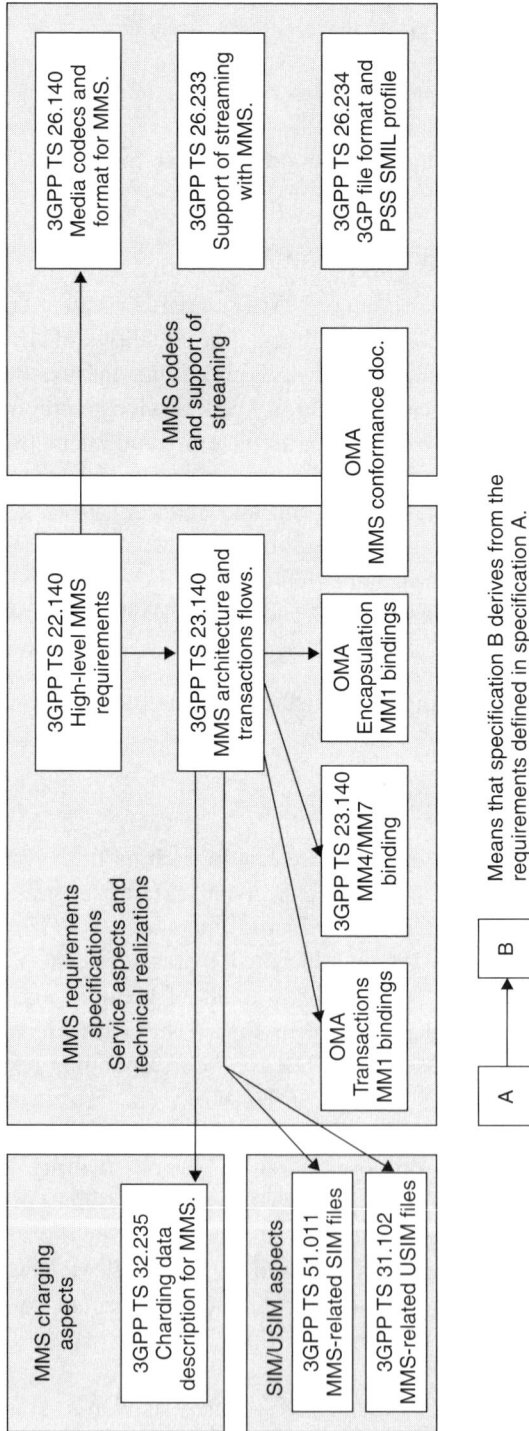

**Figure 2.8** MMS standard sets

**Table 2.4**   MMS features

| 3GPP standards | WAP forum OMA standards | Features |
|---|---|---|
| MMS Release 99 | MMS 1.0 | Basic features:<br>• Message notification<br>• Message sending/retrieval<br>• Delivery and read reports<br>• Address hiding<br>• Definition of the MM1 interface<br>• WAP configuration: WSP and HTTP. |
| MMS Release 4 | MMS 1.1 | Additional features:<br>• Reply charging<br>• Forward from notification<br>• Enhanced read report management<br>• Message sending/retrieval over secure connections<br>• Support of MMS settings and notifications in (U)SIM<br>• Definition of the MM4 interface/Updates of MM1 interface<br>• WAP stack configurations: WSP and HTTP. |
| MMS Release 5 | MMS 1.2 | Additional features:<br>• Persistent network-based storage of message (MMBox).<br>• Detailed message notification<br>• Message distribution indicator<br>• Definition of the MM7 interface/Updates of MM1 and MM4 interfaces<br>• WAP stack configurations: WSP and HTTP. |

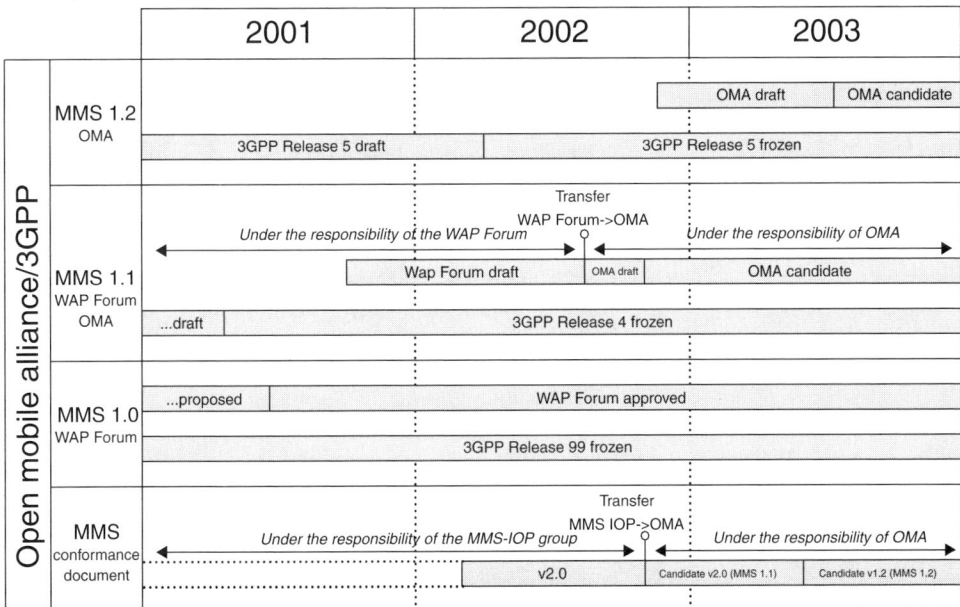

**Figure 2.9**   Availability of MMS standards

**Table 2.5**  MMS standards

| Responsible standard org. | Available from | Description |
|---|---|---|
| 3GPP | Release 99 (MMS 1.0) | Title: *MMS service aspects* [3GPP-22.140] This standard provides a set of requirements to be supported for the provision of MMS, seen primarily from the end-users' and service providers' points of view (stage 1). |
| 3GPP | Release 99 (MMS 1.0) | Title: *MMS functional description* [3GPP-23.140] This standard identifies the functional capabilities, the architecture and information flows needed to support MMS (stage 2). It also provides the details for technical realizations of several interfaces (stage 3 – MM4 and MM7). |
| WAP Forum/OMA | MMS 1.0 (Release 99) | Title: *MMS architecture overview* [WAP-205; OMA-MMS-Arch] This document is an informative document explaining the overall MMS architecture. [WAP-205] covers MMS 1.0, whereas [OMA-MMS-Arch] covers MMS 1.1 and 1.2. |
| WAP Forum/OMA | MMS 1.0 (Release 99) | Title: *MMS client transactions* [WAP-206; OMA-MMS-CTR] This standard details the transaction flows between the MMS mobile device and the MMS centre in the WAP environment. [WAP-206] covers MMS 1.0, whereas [OMA-MMS-CTR] covers MMS 1.1 and 1.2. |
| WAP Forum/OMA | MMS 1.0 (Release 99) | Title: *MMS encapsulation protocol* [WAP-209; OMA-MMS-Enc] This standard defines the binary encapsulation of MMS protocol data units. [WAP-209] covers MMS 1.0, whereas [OMA-MMS-Enc] covers MMS 1.1 and 1.2. |
| WAP Forum/OMA | MMS 1.1 (Release 4) | Title: *MMS conformance document* [OMA-MMS-Conf] This document introduces necessary restrictions in terms of transport protocol, media codecs and formats to allow interoperability between MMS devices and servers. Version 2 of this document is applicable to MMS 1.1, and to some extent to MMS 1.0. After version 2, this document was not released as version 3 but as version 1.2 since it is provided as part of the MMS 1.2 enabler release. In addition, the document defines a profile for the SMIL language, known as the MMS SMIL. |
| 3GPP | Release 5 (MMS 1.2) | Title: *MMS media formats and codecs* [3GPP-26.140] This standard represents the 3GPP recommendations regarding the applicability of media types, formats and codecs for the MMS service. Prior to Release 5, these recommendations were covered in 3GPP TS 23.140 (Release 99 and Release 4). |

**Table 2.5**   (*continued*)

| Responsible standard org. | Available from | Description |
|---|---|---|
| 3GPP | Release 4 | Title: *Streaming service: general description* [3GPP-26.233] This standard provides a general description on how media objects composing a multimedia message can be retrieved via a streaming service in the context of MMS. |
| 3GPP | Release 4 | Title: *Streaming service: protocol and codecs* [3GPP-26.234] This standard indicates which protocols and codecs are to be used for the streaming service in the context of MMS. It also defines the file structure (.3GP) for transporting video objects in multimedia messages and a SMIL profile for future MMS devices. |
| 3GPP | Release 4 | Title: *Charging data description for application services* [3GPP-32.235] This standard defines all MMS-related Charging Data Records (CDR). CDRs are required by billing systems for the generation of subscriber invoices. |
| 3GPP | Release 4 | Title: *Characteristics of the USIM application* [3GPP-31.102] This standard defines MMS-related USIM elementary files. A Release 99 version exists but does not cover MMS aspects. |
| 3GPP | Release 4 only | Title: *SIM-Mobile equipment interface* [3GPP-51.011] This standard defines MMS-related SIM elementary files. This standard is only available in Release 4 (no SIM standards after Release 4). |

# Further Reading

J. Huber, D. Weiler and H. Brand, UMTS, the mobile multimedia vision for IMT-2000: a focus on standardization, *IEEE Communications Magazine*, September, 2000.

P. Loshin, *Essential Email Standards*, John Wiley & Sons, Chichester, 1999. 3GPP TR 21.900, 3GPP working methods.

WAP work processes, WAP Forum, December, 2000 (WAP-181-TAWP-20001213-a). RFC 2026, The Internet standards process – Revision 3, IETF, October, 1996.

M.L. Olsson and J. Hjelm, OMA – changing the mobile standards game, *Ericsson Review*, **1**, 2003, 12–17.

F. Hillebrand (Editor), *GSM and UMTS: The Creation of Global Mobile Communication*, John Wiley & Sons, Chichester, 2001.

# 3

# Service Architecture

Before going deeper into the description of features offered by the Multimedia Messaging Service (MMS), it is important to understand the role of each element composing the MMS architecture. This architecture encompasses network elements required for managing MMS devices and routing multimedia messages according to user or service provider instructions in a multi-vendor environment. Network elements communicate over a set of eight identified interfaces. Interaction protocols for several of them have been standardized to ensure maximum interoperability while others are unfortunately still the subject of proprietary implementations.

One of the key interfaces in the MMS architecture enables communications between the MMS phone and the network element in charge of handling all message transactions. Available realizations of this interface are based on the Wireless Application Protocol (WAP) framework for optimal transfer of messages over bandwidth-limited radio links. Alternative realizations are being designed as outlined in Chapter 9.

This chapter presents the MMS architecture and the role of its components. In addition, an overview of the WAP framework is provided and possible realizations of the WAP-based interfaces are explained. An in-depth technical description of transactions that can occur over several of the MMS interfaces is provided in Chapter 6.

## 3.1 MMS Architecture

The MMS architecture comprises the software messaging application in the MMS phone. This application is required for the composition, sending and retrieval of multimedia messages. In addition, other elements in the network infrastructure are required to route messages, to adapt the content of messages to the capabilities of receiving devices, and so on. Figure 3.1 shows the general architecture of elements required for the realization of the MMS service.

The MMS client (also known as MMS user agent in the 3GPP terminology) is the software application shipped with the mobile handset, which allows the composition, viewing, sending, retrieval of multimedia messages and the management of reports. For the exchange of a multimedia message, the MMS client that generates and sends the multimedia message is known as the *originator MMS client*, whereas the MMS

*Multimedia Messaging Service: An Engineering Approach to MMS*   Gwenaël Le Bodic
© 2003 John Wiley & Sons, Ltd   ISBN: 0-470-86253-X

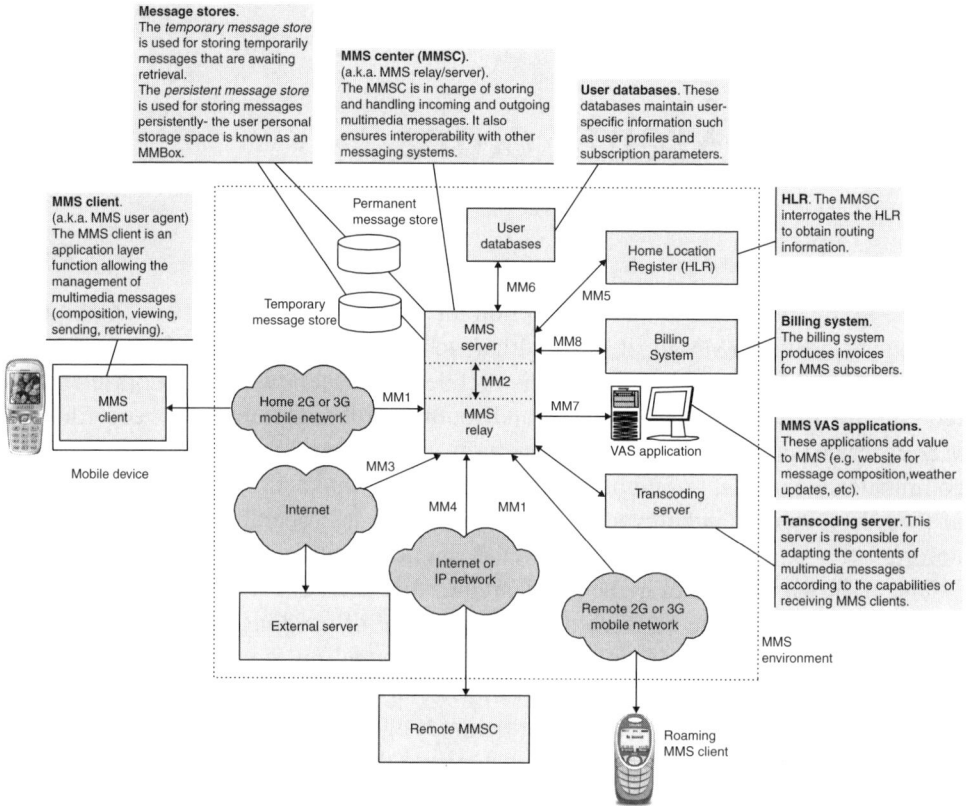

**Figure 3.1**   MMS architecture

client that receives the multimedia message is known as the *recipient MMS client*. The MMS Environment (MMSE) refers to the set of MMS elements, under the control of a single administration (MMS provider), in charge of providing the service to MMS subscribers. Recipient and originator MMS clients are attached respectively to the recipient and originator MMSEs. A key element in the MMS architecture is the MMS Centre (MMSC). The MMSC is composed of an MMS relay and an MMS server. The relay is responsible for routing messages not only within the MMSE but also outside the MMSE, whereas the server is in charge of storing messages.

## 3.2  MMS Interfaces

In an MMSE, network elements communicate via a set of interfaces. Each interface supports a number of transactions such as message submission, message retrieval and message forwarding. Each operation is associated with a set of protocol data units with corresponding parameters (e.g. recipient address, message priority, etc.). Several interfaces have been standardized in order to ensure interoperability between devices produced by various manufacturers. Other interfaces have yet to be standardized and are therefore the subject of proprietary implementations. In this book, interfaces are referred to according to the 3GPP naming convention (MM1, MM2, etc.).

- The *MM1 interface* is a key interface in the MMS environment. It allows interactions between the MMS client, hosted in the mobile device, and the MMSC. Transactions such as message submission and message retrieval can be invoked over this interface. 3GPP has defined the functional requirements of this interface. On the basis of these requirements, the WAP Forum has designed associated initial WAP-based MM1 technical realizations for the WAP environment. The Open Mobile Alliance (OMA) is now in charge of maintaining existing technical specifications for existing MM1 realizations. In addition, OMA is responsible for the development of new MM1 technical realizations for the WAP environment according to high-level requirements defined by 3GPP. This interface is also known as the $MMS_M$ interface in the WAP/OMA standards.
- The *MM2 interface* is the interface between the two internal elements composing the MMSC: the MMS server and the MMS relay. Most commercial solutions offer a combined relay and server in the form of an MMSC. Consequently, the interface between the two components is developed in a proprietary fashion. At the time of writing, no technical realization of this interface had been standardized, and it is unlikely that one would ever be standardized. This interface is also known as the $MMS_S$ interface in the WAP/OMA standards.
- The *MM3 interface* is the interface between an MMSC and external servers. Transactions invoked over this interface allow the exchange of messages between MMSCs and external servers such as email servers and SMS Centres (SMSCs). This interface is typically based on existing IP protocols. MMS standards do not specify exactly how systems should be interconnected, and it is therefore common to adapt this interface to the way the external messaging system already communicates (e.g. Simple Mail Transfer Protocol for email). This interface is also known as the E or L interface[1] in the WAP/OMA standards.
- The *MM4 interface* is the interface between two MMSCs. This interface is necessary for exchanging multimedia messages between distinct MMS environments (e.g. between two distinct mobile networks). 3GPP has standardized this interface in the Release 4 timeframe. Transactions invoked over this interface are carried out over the Simple Mail Transfer Protocol (SMTP). This interface is also known as the $MMS_R$ interface in the WAP/OMA standards.
- The *MM5 interface* enables interactions between the MMSC and other network elements. For instance, an MMSC can request routing information from the Home Location Register (HLR) or from a Domain Name Server (DNS).
- The *MM6 interface* allows interactions between the MMSC and user databases (e.g. presence server). Unfortunately, the MM6 interface is yet to be standardized.

---

[1] E stands for E-mail server and L stands for Legacy mobile messaging server.

- The *MM7 interface* fits between the MMSC and external Value Added Service (VAS) applications. This interface allows a VAS application to request services from the MMSC (message submission, etc.) and to obtain messages from remote MMS clients. Prior to 2003, implementations of this interface were all proprietary. 3GPP completed the work on the MM7 interface in the Release 5 timeframe, and commercial implementations of the standardized interface are about to appear on the market.
- The *MM8 interface* enables interactions between the MMSC and a billing system. 3GPP has standardized Charging Data Records (CDR) that are generated by the MMSC on the occurrence of certain events (e.g. message submission, message retrieval, etc.). Unfortunately, the interface used for the transfer of CDRs from the MMSC to the billing system has not been standardized yet.

## 3.3 MMS Client

The MMS client is the software application that resides in MMS-enabled mobile devices and which offers the following features:

- *Management of message, notification and reports*: Devices are commonly shipped with a unified message box for the management of MMS elements (messages, notifications and reports) and other elements such as SMS/EMS messages, WAP push messages, and so on.
- *Message composition*: The message composer is used for creating new multimedia messages.
- *Message viewing*: The message viewer is used to render received messages or to preview newly created messages before sending.
- *Configuration* of MMS preferences and connectivity parameters.
- Handling of a *remote message box* stored in the user personal network-based storage space. Such storage space is known as a Multimedia Message Box (MMBox). The support of an MMBox is optional.

## 3.4 MMS Centre

The MMS Centre (MMSC)[2] is a key element in the MMS architecture. The MMSC is responsible for handling transactions from MMS phones and transactions from other messaging systems (e.g. other MMS systems, email systems, etc.). The server is also in charge of temporarily storing messages that are awaiting retrieval from recipient MMS clients. Optionally, the server may also support a persistent message store where users can store messages persistently in their MMBoxes. This feature is particularly useful when devices have limited storage capabilities.

---

[2] The MMSC is also known as the MMS Proxy/Relay (WAP/OMA standards) or the MMS Relay/Server (3GPP standards).

The MMSC may rely on the WAP content negotiation mechanism to adapt multimedia messages to the capabilities of receiving devices. For this purpose, the MMSC has built-in transcoding capabilities or is connected to an external transcoding server as shown in Figure 3.1. The transport protocol for the interface connecting the MMSC and an external transcoding server has not yet been standardized. However, OMA is carrying some work on the definition of such an interface.

## 3.5 Wireless Application Protocol

With available MMS implementations, all MMS devices communicate with the network over the Wireless Application Protocol. WAP is the result of a collaborative work between many wireless industry players, carried out in the scope of the WAP Forum. The forum, launched in 1997 by Nokia, Phone.Com (now Openwave), Motorola and Ericsson, produced technical specifications enabling the support of applications over various wireless platforms (GSM, GPRS, UMTS, etc.). For this purpose, the WAP Forum identified and defined a set of protocols and content formats according to the standardization process presented in Chapter 2. In 2002, activities of the WAP Forum were transferred to another standardization organization: the Open Mobile Alliance.

### 3.5.1 Introduction to WAP

The WAP technology is an enabler for building applications that run seamlessly over various wireless platforms. The objective of the WAP Forum is to provide a framework for the development of applications with a focus on the following aspects:

- *Interoperability*: Applications developed by various parties and hosted on devices, produced by different manufacturers, interoperate in a satisfactory manner.
- *Scalability*: Mobile network operators are able to scale services to subscribers' needs.
- *Efficiency*: The framework offers a quality of service suited to the capabilities of underlying wireless networks.
- *Reliability*: The framework represents a stable platform for deploying services.
- *Security*: The framework ensures that user data can be safely transmitted over a serving mobile network, which may not always be the home network. This includes the protection of services and devices and the confidentiality of subscriber data.

In line with these considerations, the WAP technology provides an application model close to the World Wide Web model (also known as the web model). In the web model, content is represented using standard description formats. Additionally, applications known as web browsers retrieve the available content using standard transport protocols. The web model includes the following key elements:

- *Standard naming model*: Objects available over the web are uniquely identified by Uniform Resource Identifiers (URI).

- *Content type*: Objects available on the web are typed. Consequently, web browsers can correctly identify the type to which a specific content belongs.
- *Standard content format*: Web browsers support a number of standard content formats such as the HyperText Markup Language (HTML).
- *Standard protocols*: Web browsers also support a number of standard protocols for accessing content on the web. This includes the widely used HyperText Transfer Protocol (HTTP).

The WAP model borrows a lot from the successful web model. However, the web model, as it is, does not efficiently cope with constraints of today's mobile networks and devices. To cope with these constraints, the WAP model leverages the web model by adding the following improvements:

- The *push technology* allows content to be pushed directly from the server to the mobile device without any prior explicit request from the user.
- The adaptation of content to the capabilities of WAP devices relies on a mechanism known as the *User Agent Profile* (UAProf).
- The support of advanced *telephony features* by applications, such as the handling of calls (establishment and release of calls, placing a call on hold or redirecting the call to another user, etc.).
- The *External Functionality Interface* (*EFI*) allows 'plug-in' modules to be added to browsers and applications hosted in WAP devices in order to increase their overall capabilities.
- The *persistent storage* allows users to organize, access, store and retrieve content from/to remote locations.
- The *Multimedia Messaging Service* (*MMS*) is a significant added value of the WAP model over the web model. It relies on generic WAP mechanisms such as the push technology and the UAProf to offer a sophisticated multimedia messaging service to mobile users.

The WAP model uses the standard naming model and content types defined in the web model. In addition, the WAP model includes the following:

- *Standard content formats*: Browsers in the WAP environment, known as micro-browsers, support a number of standard content formats/languages including the Wireless Markup Language (WML) and the Extensible HTML (XHTML). WML and XHTML are both applications of the Extensible Markup Language (XML). See Box 3.1 for a description of markup languages for WAP-enabled devices.
- *Standard protocols*: Microbrowsers communicate according to protocols that have been optimized for mobile networks, including the Wireless Session Protocol (WSP) and HTTP from the web model.

---

**BOX 3.1   Markup languages for WAP-enabled devices**

The Hypertext Markup Language (HTML) is the content format commonly used in the World Wide Web. HTML enables a visual presentation of information (text, images, hyperlinks, etc.) on large screens of desktop computers. Extensible Markup Language (XML) is another markup language that is generic enough to represent the basis for the definition of many other dedicated languages. Several markup languages supported by WAP-enabled devices are derived from XML. This is the case of WML and XHTML. The WML has been optimized for rendering information on mobile devices with limited rendering capabilities. The Extensible Hypertext Markup Language (XHTML) is an XML reformulation of HTML. Both WML and XHTML are extensible since the formats allow the addition of new markup tags to meet changing needs.

---

The first WAP technical specifications were made public in 1998 and have since evolved to allow the development of more advanced services. The major milestones for WAP technology were reflected in the availability of what the WAP Forum called 'specifications suites'. Each specification suite contains a set of WAP technical specifications providing a specific level of features as shown in Table 3.1.

With WAP specification suites 1.x, the WAP device communicates with a web server via a WAP gateway. Communications between the WAP device and the WAP gateway is performed over WSP. In addition, WAP specification suite 2.x allows a better convergence of wireless and Internet technologies by promoting the use of standard protocols from the web model.

### 3.5.2 WAP Architecture

Figure 3.2 shows the components of a generic WAP architecture. The WAP device can communicate with remote servers directly or via a number of intermediary proxies. These proxies may belong to the mobile network operator or alternatively to service providers. The primary function of proxies is to optimize the transport of content from servers to WAP devices.

Supporting servers, as defined by the WAP Forum, include Public Key Infrastructure (PKI) portals, content adaptation servers and provisioning servers.

### 3.5.3 Push Technology

In a typical client/server model, a client retrieves the selected information from a server by explicitly requesting the download of information from the server. This retrieval method is also known as the *pull technology* since the client pulls some data from a server. Internet browsing is an example of models based on pull technology.

**Table 3.1**  WAP Forum specification suites

| WAP Forum specification suite | Delivery date | Description |
| --- | --- | --- |
| WAP 1.0 | April 1998 | Basic WAP framework<br>Almost no available commercial solutions since the published standards did not allow the design of interoperable solutions. |
| WAP 1.1 | June 1999 | First commercial solutions supporting:<br>• Wireless Application Environment (WAE)<br>• Wireless Session Protocol (WSP)<br>• Wireless Transaction Protocol (WTP)<br>• Wireless Markup Language (WML)<br>• WML script |
| WAP 1.2 | Nov. 1999 | Additional features:<br>• Push<br>• User Agent Profile (UAProf)<br>• Wireless Telephony Application (WTA)<br>• Wireless Identity Module (WIM)<br>• Public Key Infrastructure (PKI) |
| WAP 1.2.1 | June 2000 | Minor corrections |
| WAP 2.0 | July 2001 | Convergence with Internet technologies.<br>Additional features:<br>• Support of MMS 1.0 (3GPP Release 99)<br>• HTTP, TCP, persistent storage<br>• XHTML, SyncML, client provisioning, etc. |

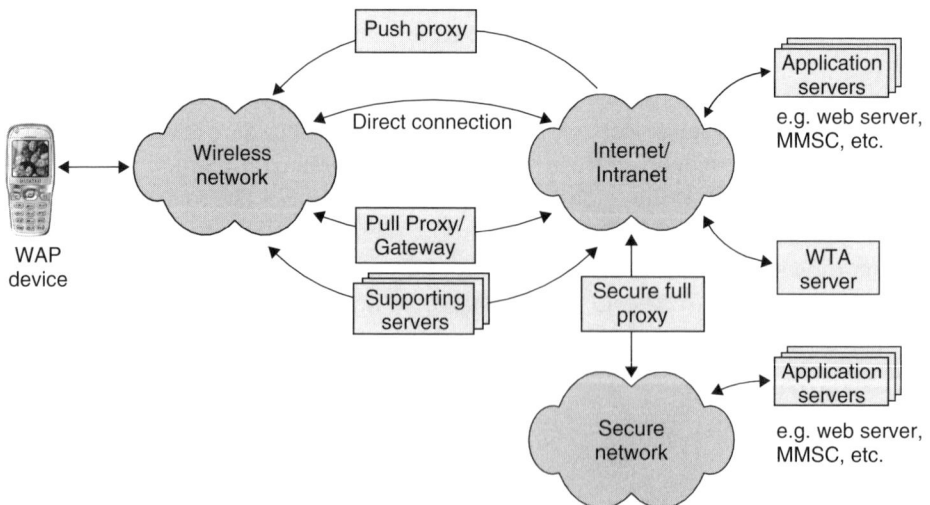

**Figure 3.2**  Generic WAP architecture

In contrast, another technology has been introduced in the WAP model and is known as the push technology. With push technology, a server is able to push some data to the WAP device with no prior explicit request from the client. In other words, the pull of information is always initiated by the client, whereas the push of information is always initiated by the server.

The push framework, defined by the WAP Forum in [WAP-250], is shown in Figure 3.3. In the push framework, a push transaction is initiated by a *push initiator*. The push initiator, usually a web server or an MMSC, transmits the content to be pushed along with delivery instructions (formatted with XML) to a *Push Proxy Gateway* (*PPG*). The PPG then delivers the push content to the WAP device according to the delivery instructions. The Push Initiator interacts with the Push Proxy Gateway using the Push Access Protocol (PAP). On the other side, the PPG uses the Push Over-The-Air (OTA) protocol (based on WSP or HTTP) to deliver the push content to the WAP device.

The Push Proxy Gateway may implement network-access-control policies indicating whether push initiators are allowed to push content to WAP devices. The Push Proxy Gateway can send back a notification to the Push Initiator to indicate the status of a push request (delivered, cancelled, expired, etc.).

Three types of browsing content can be pushed to the WAP microbrowser: Service Indication (SI), Service Loading (SL) and Cache Operation (CO). Push SI provides the ability to push content to users to notify them about electronic mail messages awaiting retrieval, news headlines, commercial offers, and so on. In its simplest form, a push SI contains a short message along with an URI. Upon receipt of the push SI, the message is presented to the user who is given the possibility of starting the service (retrieve the content) to which the URI refers. The subscriber may decide to start the service immediately or to postpone it. In contrast to push SI, push SL provides the ability to push some content to the WAP device without user explicit request. A push SL contains an URI that refers to the push content. Upon receipt of the push SL, the push content is automatically fetched by the WAP device and is presented to the user. Push CO provides a means for invalidating objects stored in the WAP device's cache memory.

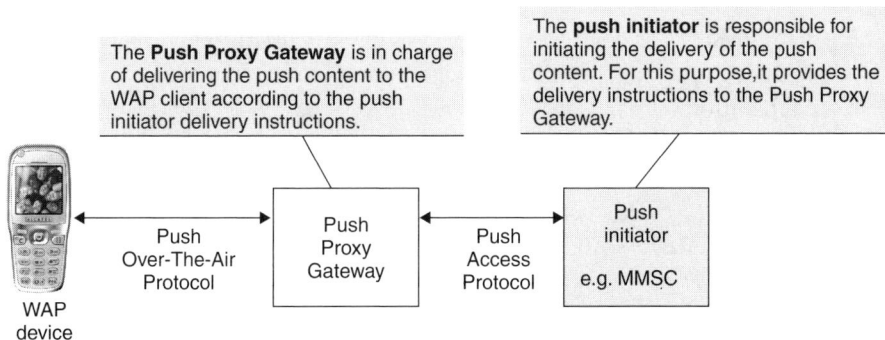

**Figure 3.3** The push framework

In addition to browsing specific push contents, information can also be pushed to other WAP-based applications such as the WTA agent, the provisioning agent and, of course, the embedded MMS client (e.g. push of message notifications and reports).

### 3.5.4 User Agent Profile

The User Agent Profile (UAProf) specification was first published in the WAP 1.2 specification suite, improved in WAP 2.0 and further enhanced recently by OMA. The objective of this specification is to define a method for describing the capabilities of clients and the preferences of subscribers. In practice, this description (known as a user agent profile) is mainly used for adapting available content to the rendering capabilities of WAP devices. For this purpose, the user agent profile is formatted using a Resource Description Framework (RDF) schema in accordance with Composite Capability/Preference Profiles (CC/PP). The CC/PP specification defines a high-level framework for exchanging and describing capability and preference information using RDF. Both RDF and CC/PP specifications have been published by W3C. The UAProf, as defined by the WAP Forum and updated by OMA in [OMA-UAProf] (version 2.0), allows the exchange of user agent profiles, also known as Capability and Preference Information (CPI), between the WAP device, intermediate network points and the origin server (web server or MMSC). These intermediate network points and origin servers (e.g. MMSC) can use the CPI to tailor the content of WSP/HTTP responses to the capabilities of receiving WAP devices. The UAProf specification defines a set of components that WAP-enabled devices can convey within the CPI. Each component is itself composed of a set of attributes or properties. Alternatively, a component can contain a URI pointing to a document describing the capabilities of the client. Such a document is stored on a server known as a profile repository (usually managed by device manufacturers or by software companies developing WAP microbrowsers). The UAProf is composed of the following components:

- *Hardware platform*: This component gathers a set of properties indicating the hardware capabilities of a device (screen size, etc.).
- *Software platform*: This component groups a set of properties indicating the software capabilities of a device (operating system, supported image formats, etc.).
- *Browser user agent*: This component gathers properties characterizing the Internet browser capabilities.
- *Network characteristics*: This component informs on network and environment characteristics such as the bearer capacity.
- *WAP characteristics*: This component advertises the device's WAP browsing capabilities. This includes information on the configuration of the WML browser and so on.
- *Push characteristics*: This component indicates the device's push capabilities. This includes the set of supported content types, the maximum message size that can be handled and whether or not the device can buffer push messages.

- *MMS characteristics*: This component describes the device capabilities for retrieving and rendering multimedia messages (MMS version, maximum message size, supported content types, etc.).

For a configuration involving a WAP device and a gateway communicating with WSP, RDF descriptions can be encoded in binary with the WAP binary XML (WBXML). In this context, the CPI is provided by the WAP device as part of the WSP session establishment request. The WAP device can also update its CPI at any time during an active WSP session. Note that the WAP gateway may also override a CPI provided by a device.

The use of UAProf in the context of MMS is further explained in Section 4.11.

### 3.5.5 WAP 1.x Legacy Configuration

With the objective of fulfilling the requirements of various services in heterogeneous mobile networks, several network configurations can coexist in the WAP environment. This section and the two following sections present the three most common configurations of the WAP environment: WAP 1.x legacy configuration, WAP HTTP proxy with wireless profiled TCP and HTTP and direct access.

Figure 3.4 shows the protocol stack of the configuration defined in the WAP specification suite 1.x[3]. This configuration is also supported by the WAP specification suite 2.0 in addition to other configurations. In this configuration, the WAP device communicates with a remote server via an intermediary WAP gateway. The primary function of the WAP gateway is to optimize the transport of content between the remote server

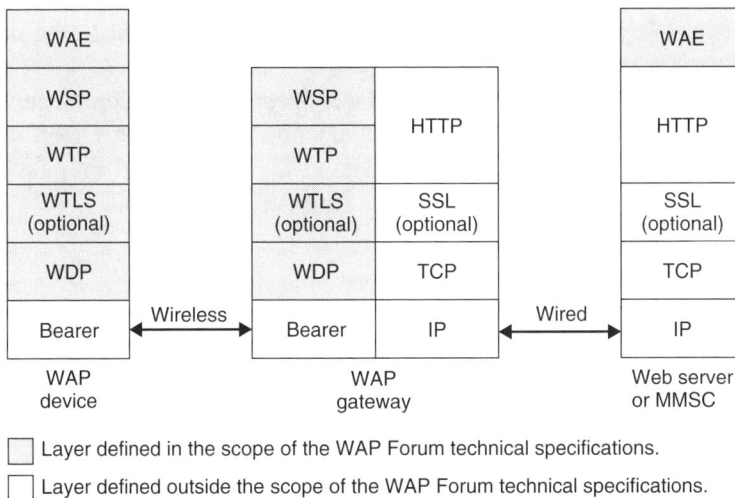

**Figure 3.4**   WAP 1.x legacy configuration with WAP gateway

---

[3] The WAP 1.x protocol stack is often referred to as the 'legacy protocol stack'.

and the WAP device. For this purpose, the content delivered by the remote server is converted into a compact binary form by the WAP gateway prior to the transfer over the wireless link. The WAP gateway converts commands conveyed between datagram-based protocols (WSP, WTP, WTLS and WDP) and protocols commonly used on the Internet (HTTP, SSL and TCP).

The *Wireless Application Environment* (*WAE*) is a general-purpose application environment in which operators and service providers can build applications (e.g. MMS client) for a wide variety of wireless platforms.

The *Wireless Session Protocol* (*WSP*) provides features also available in HTTP (requests and corresponding responses). Additionally, WSP supports long-lived sessions and the possibility to suspend and resume previously established sessions. WSP requests and corresponding responses are encoded in a binary form for transport efficiency.

The *Wireless Transaction Protocol* (*WTP*) is a lightweight transaction-oriented protocol. WTP improves the reliability over underlying datagram services by ensuring the acknowledgement and re-transmission of datagrams. WTP has no explicit connection set-up or connection release. Being a message-oriented protocol, WTP is appropriate for implementing mobile services such as browsing. Optionally, Segmentation And Reassembly (SAR) of packets composing a WTP protocol unit can be supported as described in Section 3.5.9.

The optional *Wireless Transport Layer Security* (*WTLS*) provides privacy, data integrity and authentication between applications communicating with the WAP technology. This includes the support of a secure transport service. WTLS provides operations for the establishment and the release of secure connections.

The *Wireless Data Protocol* (*WDP*) is a general datagram service based on underlying low-level bearers. WDP offers a level of service equivalent to the one offered by the Internet's User Datagram Protocol (UDP).

At the bearer level, the connection may be a circuit-switched connection (as found in GSM networks) or a packet-switched connection (as found in GPRS and UMTS networks). Alternatively, the transport of data at the bearer level may be performed over the Short Message Service or over the Cell Broadcast Service.

### 3.5.6 WAP HTTP Proxy with Wireless Profiled TCP and HTTP

Figure 3.5 shows a configuration in which the WAP device communicates with web servers via an intermediary WAP proxy. The primary role of the proxy is to optimize the transport of content between the fixed Internet and the mobile network. It also acts as a Domain Name Server (DNS) for mobile devices. With this configuration, Internet protocols are preferred against legacy WAP protocols. This is motivated by the need to support IP-based protocols in an end-to-end fashion, from the web server back to the WAP device. The protocol stack of this configuration, defined in the WAP specification suite 2.0, is shown in Figure 3.5.

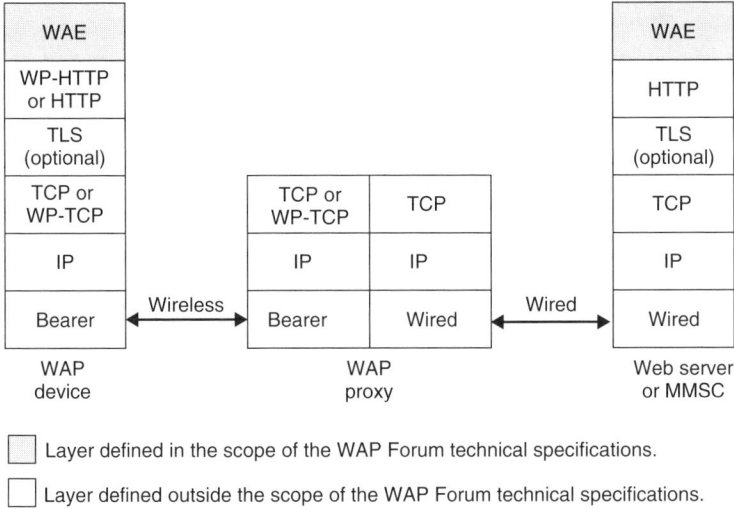

**Figure 3.5**  Configuration with WAP proxy

The *Wireless Profiled HTTP* (*WP-HTTP*) is an HTTP profile specifically designed for coping with the limitations of wireless environments. This profile is fully inter-operable with HTTP/1.1 and supports message compression.

The optional *Transport Layer Security* (*TLS*) ensures the secure transfer of content for WAP devices involved in the exchange of confidential information.

The *Wireless Profiled TCP* (*WP-TCP*) offers a connection-oriented service. It is adapted to the limitations of wireless environments but remains interoperable with existing Transmission Control Protocol (TCP) implementations.

### 3.5.7 Direct Access

Figure 3.6 shows a configuration where the WAP device is directly connected to the web server (via a wireless router that provides a bearer-level connection). The protocol stack shown in this configuration is defined in the WAP specification suite 2.0. A WAP device, compliant with the WAP 2.0 specification suite, may support all configurations by supporting WAP 1.x and WAP 2.0 protocol stacks.

### 3.5.8 WAP Configurations for MMS

In the context of MMS, the three WAP configurations introduced in the previous sections may be used. In these configurations, the MMSC plays the role of both web server and push initiator, whereas the MMS client is an application executed by the WAP mobile device.

At the time of writing, the only configuration that had been deployed was the WAP 1.x legacy configuration. A smooth transition will occur in the future towards

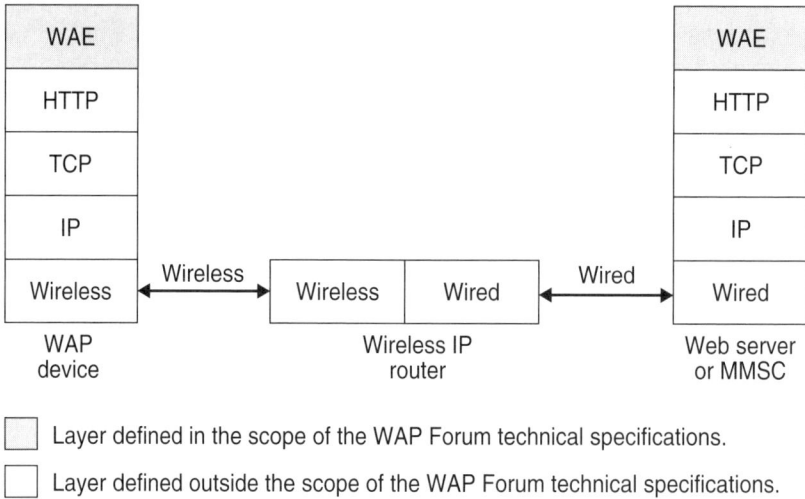

**Figure 3.6**   WAP configuration with direct access

the support of configurations without the WAP 1.x gateway. During this transition period, it is expected that operators will support multiple configurations at the same time to ensure that legacy and new devices can operate seamlessly within the same network infrastructure.

The key element of the WAP 1.x legacy configuration is the protocol stack that has been optimized for the transport of data in resource-constrained networks. A drawback of this configuration is that the amount of data that can be exchanged during a single transaction between the mobile device and the network is limited by the protocol stack specificities. For instance, the transport of a 300-kB multimedia message from the mobile device to the network cannot be accomplished in most WAP 1.x legacy environments. Configurations without the WAP 1.x gateway allow the exchange of larger amounts of data for each single transaction between the mobile device and the network, but at the cost of lower transport efficiency. As the size of messages will grow with the support of large media objects such as video clips, the migration from WAP 1.x legacy configurations to other configurations (with WAP proxy or with direct access) will become necessary for operators.

Figure 3.7 shows a configuration of the WAP environment with a WAP 1.x gateway for the support of MMS.

### 3.5.9 WTP Segmentation and Reassembly

In the WAP 1.x legacy configuration, an optional Segmentation And Reassembly (SAR) mechanism [WAP-224] allows large transactions to be segmented at the WTP level by the sender and reassembled by the receiver. SAR is specifically used when the

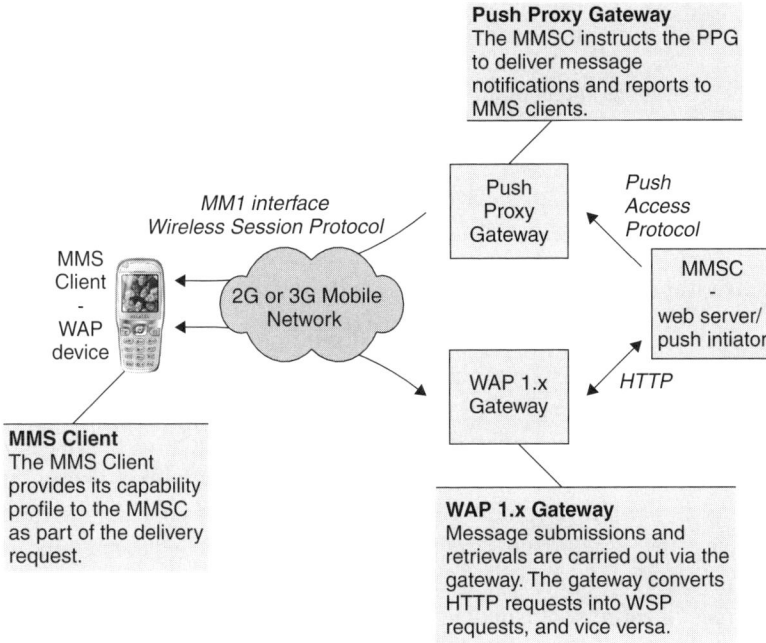

**Figure 3.7**   WAP 1.x configuration for MMS

size of a transaction (e.g. retrieval of a 50-kB message) exceeds the WTP Maximum Transmission Unit (MTU)[4]. In the context of MMS, SAR is used for transactions including the sending and retrieval of large messages. Note that, in the WAP 1.x configuration, SAR is optional and if not supported at the WTP level, then segmentation and reassembly may be supported at an underlying layer (e.g. [RFC-791] for IP, [3GPP-23.040] for SMS, etc.).

With SAR, the WTP transaction is segmented into several packets, and packets can be sent by the sender in the form of packet groups. For efficiency, the receiver acknowledges the reception of each single packet group and the sender does not start transmitting packets of a new group if the previous group has not been properly acknowledged by the receiver. A group can contain a maximum number of 256 packets. The sender determines the number of packets in a group, preferably according to the characteristics of the network and the device. The first packet group is sent without knowing the characteristics of the receiver. Therefore, the size of the first packet group should not be too large (see Box 3.2). SAR allows a selective retransmission of multiple lost packets for a given group. This feature minimizes the number of packets sent over WTP.

---

[4] The WTP Maximum Transmission Unit is the largest-size transaction that can be conveyed at the WTP level.

---

**BOX 3.2   Recommendation for maximum group size**

In order to guarantee interoperability between MMS phones and networks, it is recommended that the maximum group size for the first group shall not exceed 5120 bytes when SAR is used for conveying MMS transactions. No specific recommendation is made for packet length and number of packets in a group.

---

## 3.6 OMA Digital Rights Management

At the end of 2002, OMA published technical specifications [OMA-DRM] for mechanisms representing the basis for the management of digital rights associated with media objects downloaded via WAP download or MMS. Digital Rights Management (DRM) provides a means, for operators and providers, to control the usage of media objects once they have been downloaded to a mobile device (also known as a 'consuming device' in the DRM context). DRM enables content providers to define usage rules specifying the user's rights regarding the usage of the corresponding media object. For instance, a content provider can grant a user the rights to preview for free and charge for more sophisticated usages. Three main mechanisms are defined in OMA-DRM as shown in Figure 3.8. They differ in the way rights are communicated to the consuming device:

- *Forward lock* is the simplest of the OMA-DRM mechanisms. This is a special case of the combined delivery mechanism in which the DRM message contains only the media object, without the associated rights. For forward lock, the following set of rights applies: the user is not allowed to forward or modify the media object.
- *Combined delivery* consists of delivering the media object along with the associated rights in a single DRM message.

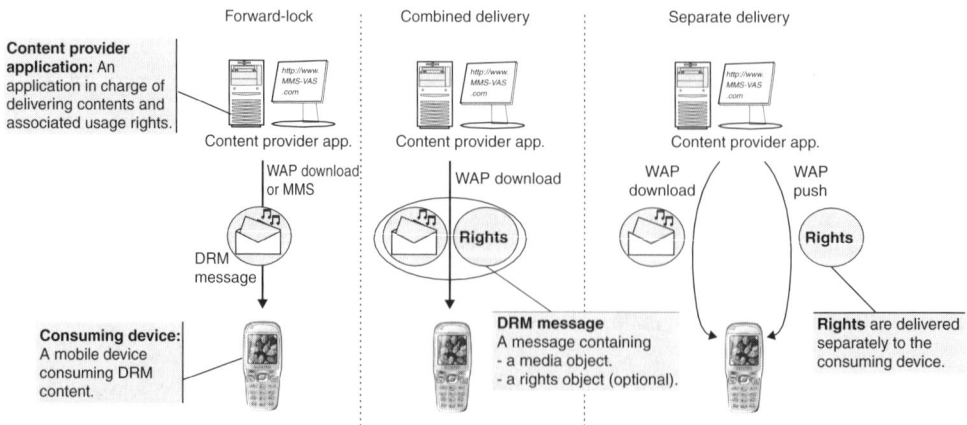

**Figure 3.8**   OMA digital rights management

- With *separate delivery*, the media object and corresponding rights are conveyed to the consuming device over separate transports. In this context, the media object is converted into a DRM Content Format (DCF) [OMA-DRM-CF]. This conversion consists of a symmetric encryption of the original media object, making the converted object unusable, unless the consuming device has the necessary Content Encryption Key (CEK) to convert the object back to its original form. The CEK along with the rights is delivered to the consuming device separately from the associated media object, typically over WAP push.

OMA DRM forward lock is of particular interest to the content-to-person scenario of MMS and is applicable from MMS 1.2. Other DRM mechanisms such as combined and separate deliveries may also become applicable to future versions of MMS.

The application of OMA DRM in the context of MMS is further explained in Section 5.6.

# 4

# Service Features

The Multimedia Messaging Service (MMS) offers several features for the support of person-to-person and content-to-person message scenarios. These features include sending and receiving multimedia messages, notifying a user that a message is awaiting retrieval, forwarding messages and managing a network-based box where messages can be persistently stored.

As for any system enabling content sharing, digital rights management (DRM) is of key importance and MMS has mechanisms for controlling the distribution of contents. In the mobile environment, devices have heterogeneous capabilities, making the provision of a homogeneous mobile messaging service difficult. To cope with this issue, MMS relies on a mechanism that adapts message contents to the capabilities of receiving devices.

Charging and billing features are also important enablers that require to be flexible enough to meet the requirements of all business models.

This chapter provides a functional-level description of all MMS features, pushing in-depth technical explanations to subsequent chapters.

## 4.1 Message Sending

One of the most basic features offered by MMS is the sending of multimedia messages. In the person-to-person scenario, message sending involves several steps as described below:

1. The user composes the multimedia message using a message composer embedded in the MMS-capable device. A multimedia message is typically structured as a multimedia slideshow (see Chapter 5). The user, also known as message originator, creates/deletes message slides and adds/removes media objects into/from message slides. Attachments can also be included in the message (e.g. electronic business cards).
2. The user instructs the mobile device to send the message to one or more recipients. According to the user's instructions, the originator MMS client (MMS management software built into the mobile device) transfers the message to the MMS Centre

*Multimedia Messaging Service: An Engineering Approach to MMS*  Gwenaël Le Bodic
© 2003 John Wiley & Sons, Ltd  ISBN: 0-470-86253-X

(MMSC) of the user's MMS Environment (MMSE). This operation is also known as *message submission*. In this context, the MMSC is known as the originator MMSC.

3. The originator MMSC performs a number of checks (message format consistency, sufficient prepaid credit, etc.). If the submission is accepted, then the originator MMSC transfers the message to the recipient MMSC(s). Note that if the message is addressed to multiple recipients, then several recipient MMSCs may be involved in the sending process (only if recipients are subscribers from MMS environments distinct from the one of the originator).

4. Upon receipt of the message, the recipient MMSC is responsible for delivering the message to the recipient MMS client as described in the next section.

For a message submission, the message originator (or originator MMS client) can assign a validity period to the message. Once this period has expired, the MMSCs automatically discard the message if it has not yet been delivered to the recipient(s). If the message originator (or originator MMS client) does not provide a validity period, then the originator MMSC provides one. Additionally, the originator MMSC may also overwrite the validity period specified by the message originator.

The basic course of message sending can also be extended with the following features:

• Request for the message to be persistently stored on the network (see also the MMBox concept explained in Section 4.9).
• Request for the originator address to be hidden from recipients (i.e. anonymous message).
• Indication that a message reply will be paid for by the message originator.
• Request for the generation of delivery and read reports.
• Indication of an earliest time of delivery for the message.

In comparison to other mobile messaging services, large messages up to tens of kilobytes can be exchanged with MMS. The sending latency perceived by the user over a General Packet Radio Service (GPRS) network can typically range from a few seconds to several tens of seconds. With most basic phone implementations, the user does not have access to other phone features while the message is being transferred to the MMSC (i.e. a modal waiting screen is displayed during the entire sending process). With more sophisticated phone implementations, message sending is performed as a background task and, during this process, the user has access to other phone features in the normal way[1].

Some early MMS devices do not support multimedia message sending. Such devices are only able to receive messages.

---

[1] With message sending in the background, resource conflicts may occur (e.g. some GPRS phones do not handle simultaneously a voice connection with the data connection required for sending the multimedia message). According to the phone capabilities, conflicting situations are resolved in various ways (e.g. voice call pre-empts the data call, message retransmissions, etc.).

## 4.2 Message Retrieval

Message retrieval consists of transferring a message from a recipient MMSC down to the local memory of an MMS device. Two retrieval modes have been designed: *immediate* and *deferred retrievals* as defined in the following sections.

The user often has the opportunity to configure the MMS device for operating in immediate retrieval mode or deferred retrieval mode. Immediate retrieval is usually the default device setting. According to the receiving device capabilities, the user may be able to indicate a specific retrieval mode to be applied when roaming.

Note that messages received and stored in an MMS device may contain media objects, which cannot be modified or redistributed (according to rights associated to the media object by the content provider/owner). In this case, the MMS device may forbid the redistribution of protected media objects contained in a multimedia message.

### 4.2.1 Immediate Retrieval

From a user viewpoint, immediate retrieval follows the Short Message Service (SMS) message retrieval model. With immediate retrieval, messages are immediately transferred, if possible, to the MMS device once they have been received by the recipient MMSC. In this way, messages appear to be immediately pushed to the MMS device without user-specific action. However, multimedia messages can be significantly larger than SMS messages. Consequently, pushing MMS messages without filtering can overload rapidly the MMS device local memory (e.g. internal flash memory). Another drawback of immediate retrieval is that it does not provide an easy way to prevent spamming (i.e. receipt of unsolicited messages). To cope with these different issues, deferred retrieval has been introduced in the MMS standards.

### 4.2.2 Deferred Retrieval

Deferred retrieval has been designed for coping with some of the limitations of immediate retrieval. Deferred retrieval consists of two successive steps:

1. Upon receipt of the multimedia message, the recipient MMSC stores the message temporarily and builds a compact notification. The notification contains information characterizing the message envelope and contents (subject, size, etc.) and is delivered to the message recipient.
2. With the notification, the recipient is informed that a message is awaiting retrieval. With deferred retrieval, it is up to the message recipient to retrieve the message at his/her own convenience. The recipient instructs the message retrieval to the MMS client, which initiates the transfer of the message from the recipient MMSC down to the MMS device local memory.

The two steps mentioned above represent the basic course of message retrieval in the deferred retrieval configuration. Alternatively, upon reception of a notification, a recipient has the opportunity to apply the following actions to the corresponding message:

- *Rejection of the message*: The message is rejected without being retrieved.
- *Forward of the message to a remote mailbox or to another recipient*: This feature is further described in Section 4.4.

Recently, MMS devices without support for message retrieval have appeared on the market. These devices are used for message sending only and are typically used as observation cameras.

In the future, it may be possible for a user to configure more sophisticated retrieval settings. For instance, the user may wish to retrieve immediately high-priority messages only, always defer the retrieval of anonymous messages, and so on.

### 4.2.3 Retrieval When Roaming

As seen in Chapter 1, the most common billing model for the person-to-person scenario relies on the sender/originator paying for the transport of the multimedia message (in both originator and recipient environments). With this model, the message is consequently free of charge for the recipient(s). However, the picture is more complex in the roaming scenario, the recipient being attached to a visited network, which is not his/her home network. In this scenario, the user may have to pay an extra fee for resources consumed on the visited network when retrieving multimedia messages. The user usually has the opportunity to set a specific retrieval mode in the case of roaming to cope with possible billing issues. A common configuration of the MMS device consists of immediately retrieving messages when attached to the home network and deferring the retrieval of messages when roaming.

### 4.2.4 Automatic Rejection of Unsolicited or Anonymous Messages

An advantage of deferred retrieval over immediate retrieval is that it offers a means for the user to check the message characteristics before the retrieval of the message. However, this selective retrieval process has to be carried out manually and this can certainly cause the user annoyance if the number of unwanted messages becomes high. To cope with this issue, devices often provide additional features for automatically rejecting multimedia messages fulfilling certain predefined criteria without user-specific intervention. Messages, which are typically rejected, are anonymous messages (originator address is not provided), advertisement messages (identified according to the message class) and messages that are not delivered via the MMSC used for submitting messages.

## 4.3 Message Reports

A message originator has the opportunity to request the generation of two types of reports for a submitted message: *delivery* and *read reports*. It is important to note that, if a message is sent to $N$ recipients, then $N$ reports of each kind may be returned back to the message originator. In order to preserve confidentiality, a recipient may deny the generation of delivery and/or read reports.

### 4.3.1 Delivery Reports

A delivery report is generated by a recipient MMSC on the occurrence of the following events:

- The message has been successfully retrieved by the recipient MMS client.
- The message validity has expired and the message has therefore been discarded without being retrieved by the recipient MMS client.
- The message has been rejected by the recipient.
- The message has been forwarded by the recipient.
- The message status is indeterminate (e.g. the message has been transferred to the Internet domain where the concept of delivery report is not fully supported).

In addition to the indication of the event, a delivery report also informs about the date when the event occurred (retrieval, deletion, forward, etc.).

### 4.3.2 Read Reports

A read report is generated by the recipient MMS client on the occurrence of the following events:

- The message has been read by the recipient.
- The message has been deleted without being read.

In addition to the indication of the event, a read report also informs about the date when the message was read or deleted without being read.

## 4.4 Message Forward

In the deferred retrieval configuration, the recipient can use the notification to request the forward of the corresponding message from the recipient MMSC to a remote mailbox (e.g. email) or to another MMS recipient. This allows a message to be transferred to another location without being retrieved first to the MMS device local memory. This is of particular interest for messages that are too large to be retrieved down to the MMS device local memory owing to memory capacity limitations.

Message forward is defined in the standards from MMS 1.1.

## 4.5 Reply Charging

During the message composition process, the message originator has the opportunity to request for reply charging to be applied to corresponding message replies. Reply charging means that the message originator is accepting to pay for a message reply per recipient. In addition, the message originator typically sets a few conditions to be honoured as a requirement for the message reply to be paid for. These conditions are as follows:

- *Reply deadline*: If the reply message is sent after the reply deadline, then the message originator is not willing to pay for it.
- *Reply size*: If the size of the reply message is larger than the reply size, then the message originator is not willing to pay for it.

If the reply charging conditions are honoured by the recipient, then the reply is paid for by the message originator and the message reply is therefore 'free of charge' for the message recipient. This applies to the first successfully delivered reply for each message recipient. If the message is forwarded by a recipient, then reply charging no longer applies to the forwarded message. Initial implementations will restrict the application of reply charging to MMS clients and Value Added Service (VAS) applications belonging to the same MMS environment. Note that at the time of writing, available commercial solutions seldom provided support for reply charging.

Reply charging is defined in the standards from MMS 1.1.

## 4.6 Addressing Modes

Three different modes are used for addressing message recipients and originators:

- *Phone number*: The phone number (e.g. +33-6-07-18-00-00), also known as a Mobile Station International ISDN Number (MSISDN), is the typical way of addressing another mobile user.
- *Email address*: The email address (e.g. gwenael@lebodic.net) is a mode used for addressing Internet users.
- *Short code*: A message can be addressed to an alphanumeric short code. A short code usually identifies a service offered by the operator or a value-added service provider in the content-to-person scenario (e.g. 'weather' or '888' typed by the user to access value-added services).

## 4.7 Settings of MMS Mobile Devices

An MMS-capable mobile device needs to be configured in order to operate properly in an MMS environment. The MMS configuration includes MMS connectivity settings and user preferences.

### 4.7.1 Connectivity Settings

MMS connectivity settings gather all parameters required to access the network infrastructure for sending and retrieving messages, reports and notifications. This includes the following parameters:

- *MMSC address*: This address is usually provided in the form of a Uniform Resource Identifier (URI) such as `http://mmsc.operator.com`.
- *WAP gateway profile*: This profile groups all parameters required to access the Wireless Application Protocol (WAP) gateway. This includes the type of WAP gateway address (e.g. IPv4, IPv6), the WAP gateway address, the access port, service type (e.g. connection-less, secured), the authentication type, authentication identification and the identification secret.
- *Bearer access parameter (e.g. GGSN)*: These parameters are required to establish a bearer-level connection with the core network. This includes the bearer type (e.g. CSD, GPRS), the type of address for the network access point (e.g. MSISDN for CSD and access point name for GPRS), bearer transfer rate, call type (e.g. analogue for CSD), the authentication type, authentication identification and the identification secret.

GPRS is the transport bearer of choice for MMS for most operators who have not yet deployed 3G infrastructures. Global System for Mobile (GSM) Circuit-Switched Data (CSD) connections are also alternative transport bearers for MMS but at the cost of higher retrieval and sending latencies. Nevertheless, the advantage of GSM circuit-switched connections is the service coverage, which is usually wider than that of GPRS. Advanced mobile devices are often able to use GPRS connections with a fallback to a GSM data connection whenever the GPRS service is not available. This ensures a high level of service availability with access to high bandwidth GPRS connections when available.

### 4.7.2 User Preferences

User preferences represent default parameter values, which are used for the creation of new messages. Users have the opportunity to update these preferences at their own convenience. Most common parameters are listed below:

- *Request for a delivery report*: Whether or not the user wishes delivery reports to be requested upon message submission.
- *Request for a read report*: Whether or not the user wishes read reports to be requested upon message submission.
- *Sender visibility*: Whether or not the user wishes his/her address to be hidden from message recipients (anonymous messages).
- *Priority of the message*: The default level of message priority (low, medium or high).
- *Time of expiry*: The time at which the message expires (e.g. 10 days after submission, one month after submission, etc.)
- *Earliest time of delivery*: The earliest time at which the message should be delivered to message recipients (e.g. one day after message submission).

Other configuration parameters could include the message-retrieval mode (deferred or immediate retrieval), the possibility to automatically reject anonymous messages and so on.

### 4.7.3 Storing and Provisioning MMS Settings

MMS settings can either be stored in the (U)SIM[2] or in the memory of the MMS phone (e.g. internal flash memory). Both methods allow the persistent storage of settings.

For the provisioning of MMS settings, three solutions are available:

1. *User configuration*: The user usually has the opportunity to configure MMS settings via the device user interface. In this scenario, settings may either be stored in the device internal memory or in the (U)SIM. Alternatively, settings may be scattered over both (U)SIM and device internal memory. If stored in the device memory, the mobile operator can instruct the device manufacturer to assign customized values to MMS settings during the device manufacturing process.

2. *Configuration of (U)SIM-stored settings by the (U)SIM issuer*: The (U)SIM issuer (usually the network operator) can update (U)SIM-stored settings and provide the (U)SIM to the user upon service subscription. Section 4.8 describes how MMS settings are stored in the (U)SIM.

3. *Over-the-Air (OTA) provisioning of settings*: OTA provisioning refers to the possibility of sending parameters dynamically over the mobile network in order to configure a mobile device remotely. This can be performed, for instance, by sending an SMS message containing MMS parameters to the mobile device. Upon reception, the device updates (U)SIM-stored or device-stored settings. Nokia and Ericsson developed a common proprietary OTA provisioning mechanism for various service settings (including MMS settings) for the provisioning of handsets[3]. However, the Open Mobile Alliance (OMA) recently published the specifications of two standardized mechanisms for the OTA provisioning of application settings in the mobile device (including MMS settings):

   • *OMA Client Provisioning (version 1.1)*: This mechanism was initially designed by the WAP Forum for the provisioning of WAP browser connectivity profiles and was further extended for the support of application parameters. In addition to WAP browsers, applications that are typically provisioned through this mechanism are email applications (POP, IMAP, SMTP), MMS clients, IMPS clients and device synchronization agents [OMA-ClientProv]. Several MMS phones available on the market already support the OMA client-provisioning mechanism for the configuration of MMS parameters.

---

[2] The term (U)SIM identifies either the Subscriber Identity Module (SIM) or the Universal SIM (USIM).

[3] Specifications for this OTA provisioning mechanism are available from the Ericsson developer website at http://www.ericsson.com/mobilityworld.

- *OMA Device Management (version 1.1.2)*:  This mechanism was initially designed by the SyncML initiative. OMA device management allows a management server to access, retrieve and update application settings stored in the mobile device [OMA-DevMan]. At the time of writing, no commercial mobile device was known to support this mechanism. However, commercial solutions supporting OMA device management are expected to appear on the market by the end of 2003.

## 4.8 Storage of MMS Settings and Notifications in the (U)SIM

As shown in the previous section, MMS settings can be stored in the SIM or USIM. The (U)SIM is provided in the form of a smart card or an electronic chip that can be inserted into a dedicated slot of GSM/GPRS mobile phones. Like MMS settings, message notifications can also be stored in the SIM or USIM. Messages are not stored in the SIM, neither are they stored in the USIM.

For this purpose, the Third Generation Partnership Project (3GPP) has defined several elementary files that can contain MMS-related data for the SIM [3GPP-51.011] and the USIM [3GPP-31.102]. As shown in Chapter 2, only Release 4 of [3GPP-51.011] defines the elementary files for the SIM, whereas similar files are defined for the USIM in [3GPP-51.011] from Release 4 onwards.

Available elementary files for MMS are listed in Table 4.1.

**Table 4.1**  MMS elementary files for (U)SIM

| Elementary file | Description |
| --- | --- |
| $EF_{MMSN}$ | Storage of a notification including<br>• notification/message status such as<br>   (a) notification read or not read,<br>   (b) corresponding message retrieved, not retrieved, rejected or<br>       forwarded,<br>• list of supported implementations (WAP-based MMS, etc.),<br>• content of the notification,<br>• pointer to a notification extension. |
| $EF_{EXT8}$ | Storage of a notification extension. |
| $EF_{MMSICP}$ | Storage of MMS issuer connectivity parameters including<br>• list of supported implementations,<br>• MMSC address,<br>• interface to core network and bearer (bearer, address, etc.),<br>• gateway parameters (address, type of address, etc.).<br>This elementary file may contain a prioritized list of connectivity parameter sets. |
| $EF_{MMSUP}$ | Storage of MMS user preferences including<br>• list of supported implementations,<br>• user preference profile name,<br>• user preference profile parameters (sender visibility, delivery report, read report, priority, time of expiry, earliest delivery time). |
| $EF_{MMSUCP}$ | Storage of MMS user connectivity parameters (same format as $EF_{MMSICP}$). |

## 4.9 Multimedia Message Boxes

From MMS 1.0, the MMSC temporarily stores messages before they are retrieved by recipient MMS devices. Once messages have been retrieved, they are stored in the device local memory (e.g. flash memory) until deleted by the recipient. Owing to limited storage capacities of mobile devices, it is often difficult to store a large number of messages in the device internal memory. Implementations from MMS 1.2 can optionally support the concept of Multimedia Message Boxes (MMBoxes). An MMBox is a network-based user message box into which messages are persistently stored. The user can access and update an MMBox remotely from the MMS device or via other means (e.g. web-user interface). With an MMBox, a user has access to the following features:

- Storing persistently submitted and retrieved messages in an MMBox (if configured or requested).
- Requesting the storage of a multimedia message that is referenced in a notification.
- Storing persistently a message that is forwarded to other recipients.
- Uploading and retrieving messages to/from an MMBox.
- Forwarding a message stored in an MMBox.
- Deleting messages stored in an MMBox.
- Viewing messages stored in an MMBox and consulting attributes associated with each message.
- Updating states and flags associated with messages stored in an MMBox.

In an MMBox, a multimedia message can be associated with one of the five mutually exclusive *states*:

- *Draft*: A message is in the draft state when it has been uploaded and stored but has not yet been submitted.
- *Sent*: A message is in the sent state when it has been stored after submission.
- *New*: A message is in the new state when it has been received by the MMSC and is persistently stored in an MMBox without having been retrieved yet.
- *Retrieved*: A message is in the retrieved state when it has been retrieved by the MMS client.
- *Forwarded*: A message is in the forwarded state when it has been forwarded by the MMS client.

In addition to the state, a message may be associated with a set of user-defined keywords (also known as *flags*). Such keywords enable the MMS client to request the list of messages complying with given keyword search criteria (e.g. 'professional', 'personal').

The user often has the opportunity to request for the MMBox *totals* (number of messages stored in the MMBox) and the MMBox *quotas* (in messages or bytes).

## 4.10 Value-added Services

Legacy mobile messaging services such as SMS and EMS (Enhanced Messaging Service) have very limited support for multimedia contents. These services fulfil the requirements of simple person-to-person scenarios well, but are often too limited for the requirements of more sophisticated content-to-person scenarios. With MMS, value-added service (VAS) providers can generate appealing multimedia contents and push such contents to multiple mobile devices as part of multimedia messages. For this purpose, 3GPP has defined a standard interface, known as the *MM7 interface*, enabling communications between the MMSC of the MMS provider (e.g. network operator) and the VAS provider (VASP) server as shown in Figure 4.1. The MM7 interface is based on the Simple Object Access Protocol (SOAP) over the HyperText Transfer Protocol (HTTP) as a transport bearer ([3GPP-23.140] from Release 5).

In this configuration, the VAS server can submit messages to the MMSC for delivery to one or multiple recipients. After submission, the VAS server has the possibility to cancel the message delivery or replace the message to be delivered if the original message has not yet been delivered. The VAS server can also receive messages from mobile users. In this context, the VAS server is identified by a dedicated service code (VAS server address). Delivery and read reports are also supported over the MM7 interface.

When submitting a message, the VAS server can indicate which party is going to pay for handling the message. The following configurations are possible:

- The VAS provider pays for the message.
- The recipient(s) pays for the message.
- The cost of handling the message is shared between the VAS provider and the recipient(s).
- Neither the recipient nor the VAS provider pays for the message.

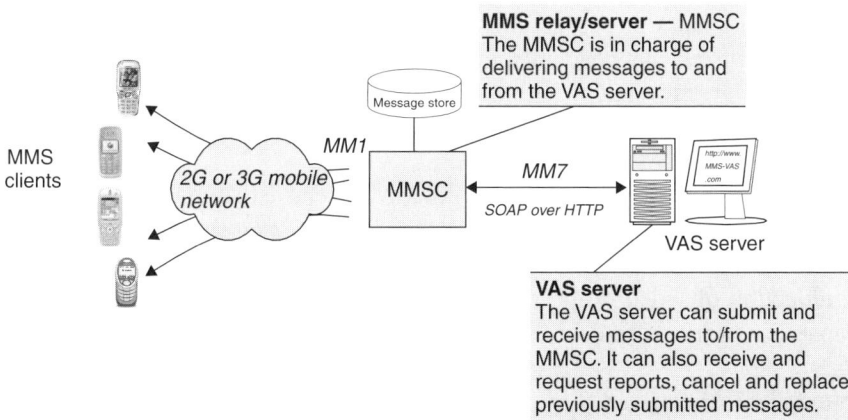

**Figure 4.1**   General configuration with VAS server

Of course, the VAS provider needs to have set up a commercial agreement with the MMS provider (e.g. network operator) in order to apply the chosen billing model for the value-added service.

A VAS provider can achieve mass distribution of information to mobile users with MMS. Furthermore, the provider can also control how message contents are redistributed using appropriate DRM mechanisms. DRM in the context of MMS is defined in the standards from MMS 1.2 [OMA-MMS-Conf].

A detailed description of the MM7 interface is provided in Chapter 6.

## 4.11 Capability Negotiation

In an environment in which mobile devices have heterogeneous capabilities, it becomes beneficial to perform adaptation of message contents according to the capabilities of receiving devices. From Release 5, 3GPP mandates the support of capability negotiation for mobile devices and highly recommends its support by MMSCs [3GPP-23.140]. In the WAP environment, the user agent profile (UAProf) technique, introduced in Section 3.5.4, is used for this purpose. Content adaptation is performed on messages before they are retrieved and is not applied to message notifications.

The recipient MMS client provides its capabilities to the MMSC prior to message retrieval. Typically, these capabilities are communicated as part of the message-retrieval request and the content-adapted message is provided as part of the retrieval response. The recipient MMS client can communicate its capabilities in three different forms:

1. Provision of the entire set of device capabilities.
2. Indication of a Uniform Resource Locator (URL) pointing to a capability profile in a remote profile repository (usually maintained by the device manufacturer). It is common for MMS providers (e.g. network operators) to copy profiles in their own internal systems to avoid having to rely on external systems but this is at the cost of additional work to update regularly phone profiles according to updates, which are applied independently in external profile repositories.
3. Provision of a differential set of information indicating changes to previously advertised capability information.

The content adaptation performed by the MMSC consists of deleting media objects that are not supported by the receiving device or adjusting the media objects to the capabilities of the receiving device (e.g. media format conversions, reduction of colour depth, reduction of video frame rate, replacement of a video clip by an image, etc.). The capability profile of a device includes the following MMS characteristics:

- Supported MMS version.
- Maximum supported message size.
- Maximum supported image resolution.

- Maximum supported colour depth.
- List of supported content types.
- List of supported character sets.
- Indication of whether or not the device supports streaming.
- List of supported Synchronized Multimedia Integration Language (SMIL) profiles (e.g. MMS SMIL, 3GPP SMIL Release 4, 3GPP SMIL Release 5).
- List of supported message content classes (e.g. text, image basic, image rich, video basic, video rich).
- Whether or not content adaptation should be disabled.

Figure 4.2 shows the process of retrieving a multimedia message when the receiving device communicates to the MMSC, a URL pointing to its capability profile in a remote repository.

As shown in Section 3.5.4, the user agent profile[4] advertises the capabilities of the device hardware and software platforms and provides the network, WAP, browser, push and MMS device characteristics. In addition to MMS-specific parameters, generic user agent profile parameters can be used by the MMSC for the content adaptation of multimedia messages. MMS-specific user agent profile parameters are defined in [OMA-MMS-CTR] and are listed in Table 4.2. Figure 4.3 shows the partial representation of the user agent profile for an MMS-capable device.

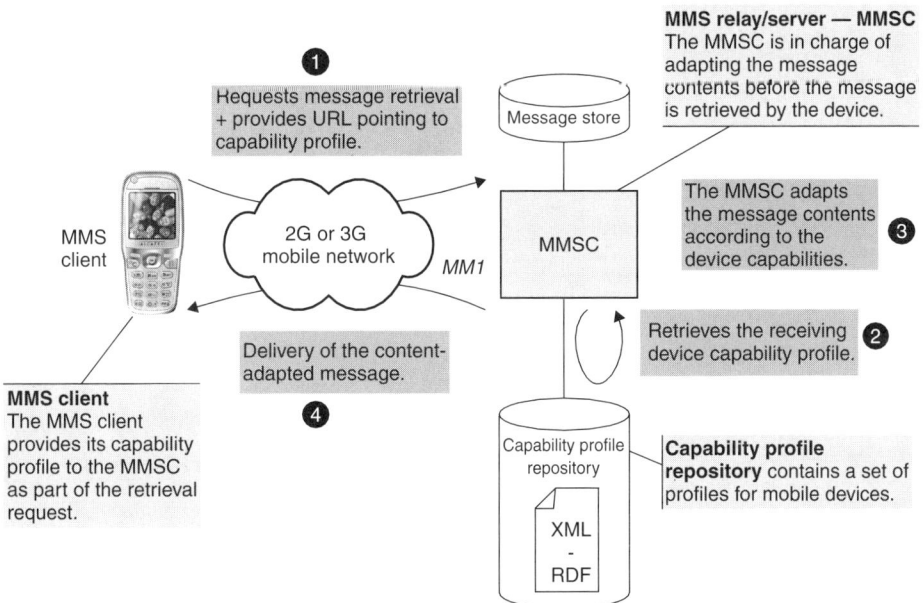

**Figure 4.2**   Content negotiation

---

[4] A list of pointers to known user agent profiles for MMS mobile devices is available from this book's companion website at http://www.lebodic.net/mms_resources.htm.

**Table 4.2**  MMS characteristics for the user agent profile as defined in [OMA-MMS-CTR] (version 1.2)

| Attribute name | Description | Examples |
|---|---|---|
| `MmsMaxMessageSize` | Maximum message size expressed in bytes. | `30720` |
| `MmsMaxImageResolution` | Maximum image dimensions expressed in pixels. | $640 \times 480$ |
| `MmsCCPPAccept` | List of supported content types. | `"image/jpeg"`, `"audio/amr"` |
| `MmsCCPPAcceptCharSet` | List of supported character sets. | `"US-ASCII"`or `"ISO-8859-1"` |
| `MmsCCPPAcceptLanguage` | List of supported languages. | `"en"`, `"fr"` for English and French respectively. |
| `MmsCCPPAcceptEncoding` | List of supported transfer encoding methods. | `"base64"`, `"quoted-printable"`. |
| `MmsVersion` | List of supported MMS version(s). | `"1.0"`, `"1.1"` |
| `MmsCCPPStreamingCapable` | Whether or not the MMS client supports streaming. | `"Yes"`or `"No"` |
| `MmsSmilBaseSet` | List of supported SMIL profiles supported by the MMS client:<br>• `SMIL-CONF-1-2` for the SMIL profile defined in [OMA-MMS-Conf] version 1.2<br>• `SMIL-3GPP-R4` for the SMIL profile defined in [3GPP-26.234] Release 4<br>• `SMIL-3GPP-R5` for the SMIL profile defined in [3GPP-26.234] Release 5 | `"SMIL-CONF-1-2"`, `"SMIL-3GPP-R4"`. |
| `MmsContentClass` | List of message content classes supported by the MMS client:<br>• `TX` for the text class<br>• `IB` for the image basic class<br>• `IR` for the image rich class<br>• `VB` for the video basic class<br>• `VR` for the video rich class | `"TX"`, `"IB"`, `"IR"`. |
| `MmsSuppressionContent Adaptation` | Request that MMSC performs no content adaptation. | `"Yes"`or `"No"` |

```
<?xml version="1.0"?>
<RDF xmlns="http://www.w3.org/1999/02/22-rdf-syntax-ns#" xmlns:rdf="http://www.w3.org/1999/02/22-rdf-
syntax-ns#" xmlns:prf="http://www.wapforum.org/UAPROF/ccppschema-20010430#">
        <rdf:Description ID="Profile">
                <prf:component>
                        <rdf:Description ID="HardwarePlatform">
                        ...
                        </rdf:Description>
                </prf:component>
                <prf:component>
                        <rdf:Description ID="SoftwarePlatform">
                        ...
                        </rdf:Description>
                </prf:component>
                <prf:component>
                        <rdf:Description ID="NetworkCharacteristics">
                        ...
                        </rdf:Description>
                </prf:component>
                <prf:component>
                        <rdf:Description ID="BrowserUA">
                        ...
                        </rdf:Description>
                </prf:component>
                <prf:component>
                        <rdf:Description ID="WapCharacteristics">
                        ...
                        </rdf:Description>
                </prf:component>
                <prf:component>
                        <rdf:Description ID="MMSCharacteristics">
                                <rdf:type resource="http://www.wapforum.org/profiles/MMS/ccppschema-
                                20010111#MMSCharacteristics"/>
                                <prf:MmsMaxMessageSize>30720</prf:MmsMaxMessageSize>
                                <prf:MmsMaxImageResolution>640×480</prf:MmsMaxImageResolution>
                                <prf:MmsVersion>1.0</prf:MmsVersion>
                                <>prf:MmsCcppAccept>
                                        <rdf:Bag>
                                                <rdf:li>image/jpg</rdf:li>
                                                <rdf:li>image/gif</rdf:li>
                                                <rdf:li>image/png</rdf:li>
                                                <rdf:li>image/bmp</rdf:li>
                                                <rdf:li>image/vnd.wap.wbmp</rdf:li>
                                                <rdf:li>application/smil</rdf:li>
                                                <rdf:li>audio/amr</rdf:li>
                                                <rdf:li>audio/midi</rdf:li>
                                                <rdf:li>text/plain</rdf:li>
                                                <rdf:li>text/x-vCard</rdf:li>
                                                <rdf:li>text/x-vCalendar</rdf:li>
                                        </rdf:Bag>
                                </prf:MmsCcppAccept>
                                <prf:MmsCcppAccept-Charset>
                                        <rdf:Bag>
                                                <rdf:li>US-ASCII</rdf:li>
                                                <rdf:li>UTF-8</rdf:li>
                                                <rdf:li>UTF-16</rdf:li>
                                        </rdf:Bag>
                                </prf:MmsCcppAccept-Charset>
                                <prf:MmsSmilBaseSet>
                                        <rdf:Bag>
                                                <rdf:li>SMIL-CONF-1-2</rdf:li>
                                        </rdf:Bag>
                                </prf:MmsSmilBaseSet>
                                <prf:MmsContentClass>
                                        <rdf:Bag>
                                                <rdf:li>TX</rdf:li>
                                                <rdf:li>IB</rdf:li>
                                        </rdf:Bag>
                                </prf:MmsContentClass>
                        </rdf:Description>
                </prf:component>
        </rdf:Description>
</RDF>
```

**Figure 4.3**   Example (partial) of user agent profile

## 4.12 Streaming

In the person-to-person scenario, the message delivery process consists of, first, notify-ing the recipient MMS client and, second, retrieving the complete multimedia message (immediate or deferred retrieval). Once the message has been retrieved, then the message content can be rendered on the recipient device upon user request. This pro-cess is suitable for most use cases. However, it is not always possible (e.g. message too large) or efficient to retrieve the complete message prior to rendering it. In these situations, it is sometimes possible to split the message content into small data chunks and to deliver the message content, chunk by chunk, to the recipient device. With this process, data chunks are directly rendered on the recipient device without wait-ing for the complete message to be retrieved. Once data chunks have been rendered by the recipient device, they are usually discarded. This means that the whole pro-cess of transmitting the message data chunks has to be carried out again, each time the MMS user wishes to read the message. This enhanced process for performing message delivery and rendering is known as the *streaming process* [3GPP-26.233; 3GPP-26.234].

### 4.12.1 Example of MMS Architecture for the Support of Streaming

In the MMS environment, it is possible to use the streaming process for the delivery of multimedia messages composed of streamable content. For this purpose, the recipient MMS client and recipient MMS environment must have streaming capabilities. Note that the support of streaming is not mandatory for the MMS client, nor is it for the MMSC. The delivery of streamable multimedia messages involves an additional element in the MMS environment. This element is known as the *media server* and is depicted in Figure 4.4. The media server may be built into the MMSC or might be a separate physical entity in the network.

The delivery of a multimedia message with the streaming process consists of five consecutive steps:

1. Upon receipt of the message, the recipient MMSC notifies the recipient MMS client that a multimedia message is awaiting retrieval. The recipient MMS client requests message retrieval (deferred or immediate). Note that this step does not differ from the normal message delivery process (without streaming support).
2. After receipt of the retrieval request, the recipient MMSC decides whether the message should be delivered in the normal way (also known as *batch delivery*) or in streaming mode. The decision is taken according to the message content, the capability negotiation and/or the user settings and preferences. If the message is to be delivered in streaming mode, then the recipient MMSC copies the message streamable content to the media server.
3. After copying the message streamable content to the media server, the recipient MMSC modifies the multimedia message prior to its delivery to the recipient

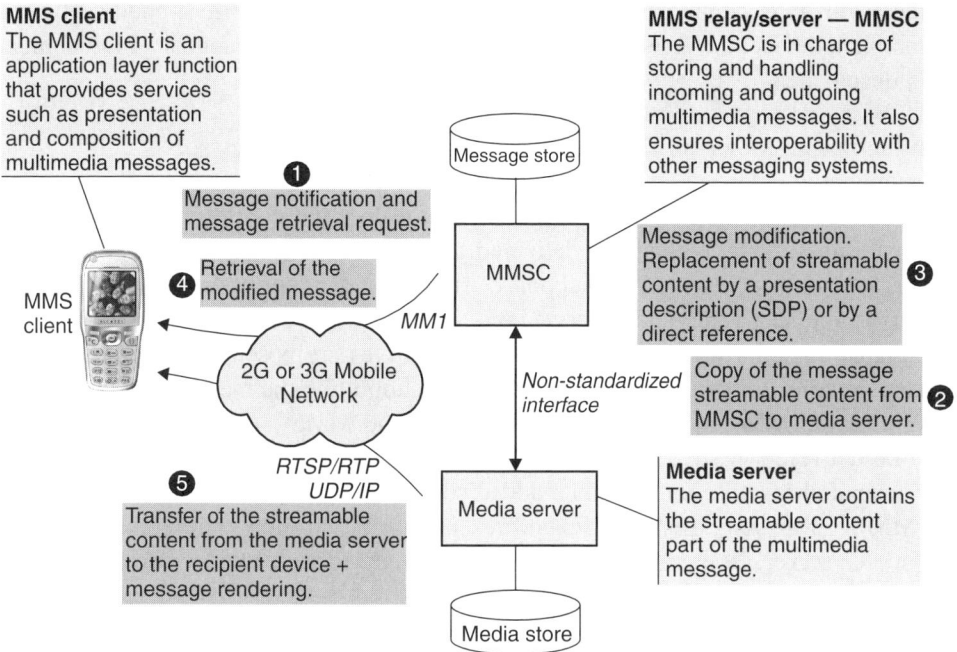

**Figure 4.4** MMS environment for the support of streaming

MMS client. The modification consists of removing the streamable content from the multimedia message and replacing it by a *presentation description* or by a *direct reference* to the content to be streamed. The presentation description indicates the streaming transport protocols to be used, the network address of the media server, the characteristics of the content to be streamed, and so on. The presentation description is formatted according to the Session Description Protocol (SDP) [RFC-2327].

4. The modified message is delivered to the recipient MMS client in the normal way and stored in the device memory.

5. At a later stage, the message recipient may decide to read the message for which a modified version is locally stored. In this situation, the recipient mobile device detects that the message contains a direct reference or a presentation description referring to a remote streamable content. In order to render the streamable content, the mobile device establishes an RTP/RTSP session with the media server and requests the Serving GPRS Server Node (SGSN) (in a GPRS environment) to activate a secondary Packet Data Protocol (PDP) context[5] with a level of quality of service fulfilling the streaming mode requirements. The recipient MMS client instructs the media server to start the delivery of the streamable content with

---

[5] In a GPRS environment, a mobile device is required to activate a Packet Data Protocol (PDP) context with the network in order to be able to transfer data to and from the network.

the RTSP Play command. Once the streamable content has been rendered by the recipient device, the RTSP/RTP session is released and the secondary PDP context is deactivated.

In the person-to-person scenario, the way the multimedia message is delivered to the recipient is independent of the way it has been initially submitted by the message originator. In other words, the message originator does not control the way (batch or stream) a multimedia message is delivered to message recipients.

This section has presented the support of streaming delivery in the person-to-person scenario. Note that value-added service providers may also send messages, containing streamable content, to mobile users. Two alternatives are possible in the content-to-person scenario. The first alternative consists of submitting the message, in the normal way, over the MM7 interface. In this situation, the MMSC decides if the message is to be delivered in streaming or batch mode (as in the person-to-person scenario). The second alternative for the value-added service provider consists of submitting to the MMSC a message that contains a presentation description or reference pointing to a media server. In this situation, the value-added service provider decides that the message is to be delivered in streaming mode.

An example of a message containing a direct reference to the content to be streamed is provided in Section 5.4.9.

### 4.12.2 Streaming Protocols: RTP and RTSP

Two protocols can be used for retrieving some streamable content from a media server: the *Real-Time Transport Protocol* (RTP) [RFC-1889] and the *Real-Time Streaming Protocol* (RTSP) [RFC-2326].

RTP is a generic transport protocol allowing the transfer of real-time data from a server. In the MMS environment, RTP enables a one-way communication from the media server to the MMS client. This communication enables the delivery of streamable content, stored in the media server, to the MMS client in streaming mode. RTP usually relies on the connection-less User Datagram Protocol (UDP) (itself relying on the IP protocol). Compared with the connection-oriented Transmission Control Protocol (TCP), UDP allows a faster and more resource-efficient transfer of data. However, UDP lacks a mechanism for reporting loss of data chunks. Consequently, with RTP, lost data chunks are not retransmitted (note that for the transfer of real-time data, transport reliability is not as important as timely delivery). RTP relies on three basic mechanisms for the transport of real-time data:

- *Sequence numbering*: UDP does not always deliver data chunks in the order they were sent by the media server. To cope with this, the media server tags data chunks with an incremental sequence number. This sequence number allows the MMS client to reorganize data chunks in the correct order. The sequence number is also used by the MMS client for detecting any loss of data chunks.

- *Time-stamping*: The media server time-stamps all data chunks prior to their delivery to the MMS client. After receiving data chunks, the MMS client reconstructs the original timing in order to render the streamable content at the appropriate rate.
- *Payload-type identifier*: This identifier indicates the encoding/compression schemes used by the media server. According to the payload-type identifier, the MMS client determines how to render the message content.

With RTSP, the MMS client can control the way the streamable content is delivered from the media server. For this purpose, the MMS client instructs the media server to start, pause and play the message content during the content delivery and rendering. In other words, with RTSP, the MMS client can control the content delivery from the media server in the same way as a person controls a VCR with a remote controller. This is why RTSP is sometimes known as a 'network remote control' for multimedia servers. RTSP is used for establishing and controlling streaming sessions but is not used for the transport of streamable content. The transport of the streamable content is handled by transport protocols such as RTP. While most streaming content delivery uses RTP as a transport protocol, RTSP is not tied to RTP.

## 4.13 Charging and Billing

MMS has been deployed worldwide and each operator has adapted its own MMS billing model(s) to the local market requirements and MMS usages (e.g. content-to-person, person to person, etc.). To enable this, 3GPP has designed a flexible charging framework for the generation and treatment of relevant charging information. The design of this framework is based on generic charging principles identified in [3GPP-32.200] and charging principles specific to MMS as identified in [3GPP-22.140].

The MMSC generates charging information in the form of Charging Data Records (CDRs) on the occurrence of specific events (message submission, message retrieval, etc.). Once generated, CDRs are transferred from the MMSC to the billing server over the MM8 interface. Note that, at the time of writing, the MM8 interface was still to be standardized. The billing system processes all CDRs in order to produce subscriber invoices.

3GPP has identified all MMS events for which charging information should be generated [3GPP-32.235]. Corresponding CDRs are categorized in five classes:

- *CDRs for the originator MMSC*: This class groups CDRs created by the originator MMSC for events over MM1 and MM4 interfaces.
- *CDRs for the recipient MMSC*: This class groups CDRs created by the recipient MMSC for events over MM1 and MM4 interfaces.
- *CDRs for the forwarding MMSC*: This class groups CDRs created by a forwarding MMSC.

- *CDRs for the MMSC supporting the MMBox concept*: This class groups CDRs created by an originator MMSC supporting MMBoxes.
- *CDRs for VAS applications*: This class groups CDRs created by the MMSC for events occurring over the MM7 interface.

Table 4.3 provides the list of events and corresponding CDR names [3GPP-32.235].

CDRs include information such as the duration of message transmission, charging information (post-paid, pre-paid), message content type, message class and priority, message size, reply charging parameters, recipient addresses and so on.

**Table 4.3**  MMS charging data records

| Event | Interface | Category | CDR name |
|---|---|---|---|
| Message submission | MM1 | Originator | O1S-CDR |
| Forward request | MM4 | Originator | O4FRq-CDR |
| Forward response | MM4 | Originator | OFRs-CDR |
| Delivery report | MM4 | Originator | O4D-CDR |
| Delivery report | MM1 | Originator | O1D-CDR |
| Read–reply report | MM4 | Originator | O4R-CDR |
| Read–reply originator | MM1 | Originator | O1R-CDR |
| Originator message deletion | n/a | Originator | OMD-CDR |
| Message forward | MM4 | Recipient | R4F-CDR |
| Notification request | MM1 | Recipient | R1NRq-CDR |
| Notification response | MM1 | Recipient | R1NRs-CDR |
| Message retrieval | MM1 | Recipient | R1Rt-CDR |
| Acknowledgement | MM1 | Recipient | R1A-CDR |
| Delivery report request | MM4 | Recipient | R4DRq-CDR |
| Delivery report response | MM4 | Recipient | R4DRs-CDR |
| Read–reply recipient | MM1 | Recipient | R1RR-CDR |
| Read–reply report request | MM4 | Recipient | R4RRq-CDR |
| Read–reply report response | MM4 | Recipient | R4RRs-CDR |
| Recipient message deletion | n/a | Recipient | RMD-CDR |
| Forwarding | n/a | Forwarding | F-CDR |
| Message store | MM1 | MMBox | Bx1S-CDR |
| Message view | MM1 | MMBox | Bx1V-CDR |
| Message upload | MM1 | MMBox | Bx1U-CDR |
| Message deletion | MM1 | MMBox | Bx1D-CDR |
| Message submission | MM7 | VAS | MM7S-CDR |
| Delivery request | MM7 | VAS | MM7DRq-CDR |
| Delivery response | MM7 | VAS | MM7DRs-CDR |
| Message cancel | MM7 | VAS | MM7C-CDR |
| Message replace | MM7 | VAS | MM7R-CDR |
| Delivery report request | MM7 | VAS | MM7DRRq-CDR |
| Delivery report response | MM7 | VAS | MM7DRRs-CDR |
| Read report request | MM7 | VAS | MM7RRq-CDR |
| Read report response | MM7 | VAS | MM7RRs-CDR |

## 4.14 Security Considerations

It is possible to establish secure connections between the MMSC and the MMS client for the retrieval or submission of multimedia messages. In the WAP environment, MMS relies on Wireless Session Protocol (WSP) or HTTP requests for the transport of messages (see Section 3.5.8). A message submission or retrieval request includes a Uniform Resource Identifier (URI), which starts with a protocol scheme such as `http` or `https`.

The `http` scheme does not imply the use of a particular transport protocol between an MMS client and an MMSC. In this context, communications can be performed over the following protocols:

- Via a WAP 1.x gateway supporting protocol conversion between WSP and HTTP.
- Wireless-profiled HTTP or HTTP.

On the other hand, the `https` scheme implies the use of a secure connection between the MMS client and the MMSC. According to the WAP configuration in place, the secure connection can be established over the following secured protocols:

- Wireless-profiled HTTP or HTTP in accordance with the WAP TLS profile and tunnelling specification [WAP-219].
- Via a WAP 1.x gateway using WSP over Wireless Transport Layer Security (WTLS) for the security layer between the MMS client and the gateway and using HTTP over TLS or SSL between the gateway and the MMSC.

It is also possible to secure communications between the MMSC and servers of value-added service providers. HTTP sessions between the VAS server and the MMSC can be established with HTTP over TLS or SSL.

Authentication techniques defined in [RFC-2617] for 'basic' and 'digest' authentication can be used to authenticate the VAS provider during the establishment of sessions for message submissions. A VAS provider may also be authenticated with a VAS identifier and a password prior to interacting with the MMSC. Note that the authentication mechanism based on public/private key cryptography can also be used in this context.

# 5

# The Multimedia Message

One of the major differentiating characteristics of the Multimedia Messaging Service (MMS) is the organization of message media objects as user-friendly multimedia presentations. This organization enables the creation of simple point-shoot-and-send messages with feature-limited device-embedded composers (person-to-person) to the design of sophisticated multimedia slideshows with feature-rich professional tools (content-to-person). The internal message structure, which has been selected for MMS, has close similarities with the structure of Internet email messages. Like most email messages, multimedia messages are structured as multipart messages containing various media objects such as texts, images, sounds, video clips and so on.

A multimedia multipart message usually contains a scene description. A scene description tells the receiving device how media objects contained in the message are choreographed on the screen and over time in order to produce a meaningful multimedia presentation.

This chapter presents the structure of the multimedia message and describes how to design scene descriptions. It also explains the MMS categorization of media types, which allow devices to interoperate in a coherent fashion.

## 5.1 Multipart Structure

In the MMS Environment (MMSE), a multimedia message can take multiple forms in order to be efficiently conveyed over the various transport bearers composing the full message transfer path. The link between an MMS client and the MMS Centre (MMSC) is often bandwidth-limited (particularly over the radio part); therefore, multimedia messages are binary-encoded for efficient transfer over this link. Alternatively, the multimedia message is text-encoded for transfers over Internet protocols between MMSCs, from an MMSC towards the Internet domain or from/to Value Added Service (VAS) servers.

Contents of a multimedia message range from simple text to sophisticated media objects with optional intermedia synchronization. Multimedia message objects are wrapped into an *envelope*, which allows various network elements to route the message towards the recipients (addresses of primary and secondary recipients) and which characterize the message contents (class, priority, subject, etc.).

*Multimedia Messaging Service: An Engineering Approach to MMS*   Gwenaël Le Bodic
© 2003 John Wiley & Sons, Ltd   ISBN: 0-470-86253-X

The basic structure of the message envelope is derived from the one defined by the Internet Engineering Task Force (IETF) for the Internet email [RFC-2822] ([RFC-2822] obsoletes the well-known [RFC-822]). Furthermore, the encapsulation method for inserting media objects into a multipart message is derived from the Multipurpose Internet Mail Extensions (MIME) standards [RFC-2045; RFC-2046; RFC-2047; RFC-2048; RFC-2049]. Media objects are encapsulated in individual object containers known as *body parts*. The WAP Forum and the Third Generation Partnership Project (3GPP) have published recommendations indicating how to use these IETF standards (RFC 2822 and MIME) in the context of MMS. Furthermore, the WAP Forum has complemented IETF standards by defining and publishing a generic binary representation of the multipart structure for a more efficient transport over bandwidth-limited bearers. The Open Mobile Alliance (OMA) is now the organization responsible for the maintenance and evolution of this binary representation [OMA-WSP] and of its MMS extensions [OMA-MMS-Enc]. A detailed description of this binary representation for MMS is provided in Section 6.2.11.

Figure 5.1 shows the multimedia message structure relying on a message envelope (message header) and message contents (body parts):

### 5.1.1 Message Envelope

As shown above, a multimedia message consists of an envelope (also known as message header) and message contents (encapsulated in message body parts). The envelope informs about the following message characteristics:

- Address of the message originator (From)
- Address of the message recipient(s), organized into primary recipients and secondary recipients (To, Cc and Bcc)

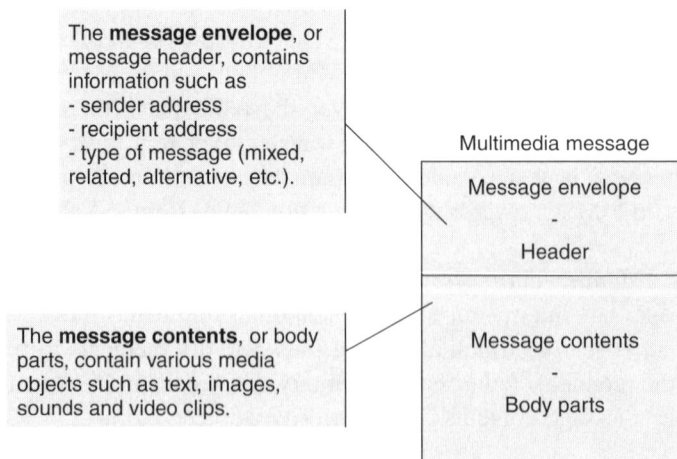

**Figure 5.1**   Structure of a multimedia message

- Priority of the message (low, medium or high)
- Class of the message (auto, personal, informational or advertisement)
- Date and time when the message was sent
- Validity period
- Reply charging parameters
- Request for delivery and/or read reports
- Message subject
- Sender visibility
- Earliest delivery time
- Message distribution indicator
- MMBox status.

Each of these characteristics has a dedicated set of parameters. Parameters are either defined in [RFC-2822] or specifically defined for MMS. In the latter case, the name of the parameter is prefixed with X-MMS. Table 5.1 shows common parameters that can be included in a message envelope/message header.[1]

The value assigned to the Content-Type parameter of the message header indicates how the message body parts are organized in the message. The three values presented in Table 5.2 can be used in the context of MMS.

Multiple parts of a message are separated by boundary delimiters. The name of a boundary delimiter is assigned to the subtype parameter called boundary. The name of a boundary delimiter has a size ranging from 1 to 70 characters. With a multipart textual representation, each part begins with two hyphens (--) followed by the boundary delimiter name. The multipart message is terminated by a carriage return/line feed followed by two hyphens, the boundary delimiter name and two additional hyphens.

In theory, multipart structures can be nested in upper multipart structures in order to build a complete hierarchy of body parts. In the context of MMS, the most common structure consists of having a main multipart structure (mixed or related) containing all body parts composing the message (images, sounds, video clips, etc.).

### 5.1.2 Encapsulation of Media Objects

In addition to the envelope, multimedia messages contain media objects such as texts, images, sounds and video clips. Each media object is encapsulated in a container known as a body part. The overall multipart structure of a multimedia message is depicted in Figure 5.2.

Each body part is associated with a number of header parameters as shown in Table 5.3.

The only mandatory parameter for a body part is the Content-type parameter.

---

[1] For the sake of clarity, parameters related to the network storage of messages (MMBox) are not presented in Table 5.1. An exhaustive list of parameters is provided in Chapter 6.

**Table 5.1**   Message envelope/header parameters

| Parameter name | Description |
| --- | --- |
| To | Address of the primary message recipient(s). |
| Cc | Address of the secondary message recipient(s). Carbon copy. |
| Bcc | Address of the secondary message recipient(s). Blind carbon copy. |
| From | Address of the message originator. |
| Date | Date and time when the message was sent. |
| Message-ID | A unique identifier for the message. This helps to correlate delivery and read reports with the original message. |
| Subject | Subject of the message. |
| X-Mms-Expiry | Validity period of the message. After this date, MMSCs can discard the message if it has not yet been delivered to the recipient. |
| X-Mms-Delivery-Time | The message originator can specify an earliest delivery time for the message. |
| X-Mms-Distribution-Indicator | A value-added service provider may use this flag to indicate that the message cannot be redistributed freely. |
| X-Mms-Reply-Charging | Whether or not the message originator has requested reply charging. |
| X-Mms-Reply-Charging-Deadline | The reply must be sent before the specified deadline for the reply to be paid for by the message originator. |
| X-Mms-Reply-Charging-Size | The size of the message must be smaller than the specified size for the reply to be paid for by the message originator. |
| X-Mms-Reply-Charging-ID | A unique identifier for the reply-charging transaction. |
| X-Mms-Delivery-Report | Whether or not the message originator requested a delivery report to be generated. |
| X-Mms-Read-Report | Whether or not the message originator requested a read report to be generated. |
| X-Mms-Message-Class | The class of the message (auto, personal, informational, advertisement). |
| X-Mms-Priority | The priority of the message (low, medium or high). |
| X-Mms-Sender-Visibility | Whether or not the message originator requested his/her address to be hidden from recipients. |
| Content-type | The content type of the message. A description of values that can be assigned to this parameter is provided in Table 5.2. |

## 5.2  Message Content Domains and Classes

As shown in previous sections, a message is composed of a header (envelope) and body parts. In theory, any media object with a proper body part encapsulation may be

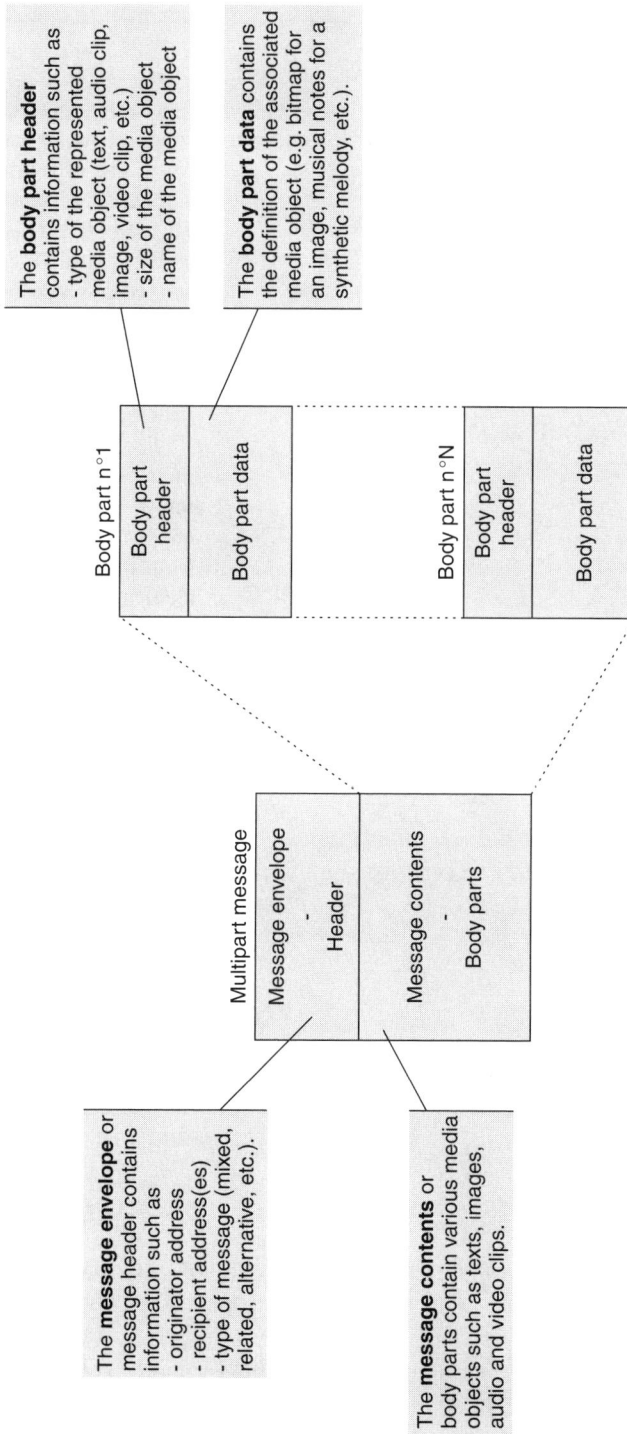

**Figure 5.2** Structure of a multipart message

The **body part header** contains information such as
- type of the represented media object (text, audio clip, image, video clip, etc.)
- size of the media object
- name of the media object

The **body part data** contains the definition of the associated media object (e.g. bitmap for an image, musical notes for a synthetic melody, etc.).

Body part n°1
Body part header
Body part data

Body part n°N
Body part header
Body part data

Multipart message
Message envelope - Header
Message contents - Body parts

The **message envelope** or message header contains information such as
- originator address
- recipient address(es)
- type of message (mixed, related, alternative, etc.).

The **message contents** or body parts contain various media objects such as texts, images, audio and video clips.

**Table 5.2**   Multipart messages/content types

| Content-type value[a] | Description |
| --- | --- |
| `Multipart/Mixed`<br>or<br>`application/vnd.wap.multipart.mixed` | A mixed multipart structure contains one or more body parts. The order in which body parts appear has no significance. This structure is commonly used when the message does not contain a scene description. |
| `Multipart/Alternative`<br>or<br>`application/vnd.wap.multipart.alternative` | An alternative multipart structure contains one or more body parts. Each body part of the structure is a representation of a single element. For instance, the structure can contain an image in the JPEG format and another one in the GIF format. Both images represent the same element. It is up to the receiving MMS client to select the most appropriate representation while rendering the message. This content type is seldom used in the context of MMS. |
| `Multipart/Related`<br>or<br>`application/vnd.wap.multipart.related` | A related multipart structure [RFC-2557] is used for aggregating multiple body parts into a single structure. The optional `Start` parameter refers to a starting body part. For instance, in a multimedia message, the `Start` parameter typically refers to a scene description in the SMIL format. |

[a] WAP registered content types are listed at http://www.wapforum.org/wina/wsp-content-type.htm.

inserted in a multimedia message. Practically, a minimum set of common formats and codecs has been identified in order to ensure interoperability between MMS phones produced by different vendors.

3GPP has made recommendations regarding the support of formats and codecs in the context of MMS ([3GPP-23.140] prior to Release 5 and [3GPP-26.140] from Release 5). However, 3GPP recommendations are not restrictive enough to ensure

**Table 5.3**  Message body part header parameters

| Parameter name | Description |
|---|---|
| Content-ID | A unique identifier for the body part in the multipart message [RFC-2392]. The identifier is typically inserted between square brackets. Example: <br> Content-ID: <ui_bp_0006> |
| Content-Location | A user-readable name that is commonly used for naming the media object contained in the body part [RFC-2557]. This is particularly useful when the user wishes to extract the media object from the message (image to be used as wallpaper, audio clip to be used as ring tone, etc.). Example: <br> Content-Location: house.jpg |
| Content-Disposition | An indication on how the media object should be displayed [RFC-1806]. Values can be INLINE for an inline display of the corresponding object or ATTACHMENT for a display in the form of an attachment. However, this parameter is seldom used in the context of MMS. Another technique is used in MMS for differentiating inline message objects from message attachments (see Section 5.4.8). |
| Content-Type | The content type of the corresponding message object. A list of well-known content types for MMS is provided in Appendix A. Examples: <br> Content-type: image/jpeg <br> Content-type: audio/midi <br> Optionally, certain content types can be associated with one or more parameters such as the character set or a name/filename. Examples: <br> Content-type: text/plain; charset = US-ASCII <br> Content-type: image/GIF; name = house.jpg |

sufficient interoperability between MMS phones. In 2001, an informal group of vendors known as the MMS interoperability group (MMS-IOP) met to design a specification restricting the number of formats and codecs supported by early MMS phones with the objective to ensure interoperability. This specification, referred to as the *MMS conformance document*, constituted the basis for the design of all commercial MMS devices available today on the market. Considering the importance of guaranteeing interoperability between MMS devices, the responsibility for the MMS conformance document was transferred in 2002 to the Open Mobile Alliance. OMA published its first MMS conformance document, part of an enabler release, as MMS conformance document version 2.0.0 [OMA-MMS-Conf] (MMS 1.1). The MMS conformance version 2.0.0 is applicable to devices compliant to MMS 1.1 and to some extent MMS 1.0. The successor version of this document was not released as version 3.0 but as version 1.2 since it is published as part of the MMS 1.2 enabler release. Version 1.2 builds up from the MMS conformance document version 2.0.0 by adding the support of video and introducing the concept of message content domains and classes. The MMS conformance document version 1.2 is not only applicable to devices compliant with MMS 1.2 but also to devices compliant with previous versions of the MMS standards (MMS 1.0 and MMS 1.1). It primarily targets devices whose commercial availability starts at the end of 2003. A new version of this document is expected to be produced later by OMA, primarily targeting devices whose commercial availability starts at the end of 2004.

---

**BOX 5.1    Support of MMS formats/codecs: 3GPP TS 26.140 or OMA MMS conformance document?**

Two MMS standards provide recommendations regarding the support of formats and codecs in the context of MMS. Which one is applicable, 3GPP TS 26.140 or the OMA MMS conformance document? Currently, the OMA MMS conformance recommends the use of a group of formats/codecs that constitute a subset of the ones identified in 3GPP TS 26.140. In other words, the MMS conformance document is more restrictive than 3GPP TS 26.140. For the sake of interoperability, the OMA MMS conformance is therefore the applicable standard for MMS 1.0, 1.1 and 1.2. The ultimate goal is to make 3GPP TS 26.140 applicable in the future. This will only be made possible if 3GPP TS 26.140 becomes more restrictive, or if mobile device capabilities improve to a stage where the application of 3GPP TS 26.140, in its current form, does not cause any interoperability issue.

---

### 5.2.1 Message Content Domains

Considering the full standardization picture, messages can be categorized into three nested message content domains [OMA-MMS-Conf] (version 1.2):

- *Unclassified message content domain*: This domain basically includes all messages composed of media objects in the unlimited set of available formats/codecs. Devices allowing the composition of messages belonging to the unclassified domain, without restrictions regarding the set of usable formats/codecs, are subject to interoperability issues at a high-risk level.
- *Standard message content domain*: This domain groups messages containing media objects with formats/codecs identified by 3GPP in [3GPP-26.140]. Regarding the number of formats/codecs in [3GPP-26.140] and the optional nature of most of them, devices allowing the composition of messages belonging to the standard domain, without restrictions other than using formats/codecs identified in [3GPP-26.140], are subject to a medium interoperability risk.

- *Core message content domain*: This domain identifies messages containing media objects with formats/codecs identified in the OMA conformance document [OMA-MMS-Conf]. Devices allowing the composition of messages belonging to the core message domain are subject to a low interoperability risk. At the network level, content adaptation may be applied on messages belonging to the standard message domain in order to convert them into messages belonging to the core message domain. Messages belonging to the core message domain can further conform to one of the five hierarchical message content classes as defined in the following section.

### 5.2.2 Message Content Classes

In the MMS conformance document version 1.2, the concept of message content classes[2] in the core message domain is introduced with the objective to guarantee interoperability between conformant MMS clients. The definition for each class identifies the maximum size for a message compliant to that class, a set of allowed formats and codecs and whether or not OMA DRM forward-lock is applicable. MMS clients that comply with the conformance document version 1.2 do support the requirements of two or more content classes in terms of message creation, submission, retrieval and/or presentation. Conformant MMS clients are all expected to support the *text class*, which is the simplest of the five defined content classes. In addition, a conformant MMS client also supports at least another content class. MMS client conformant to the version 2.0.0 of the MMS conformance document do support the *image basic* content class. In addition, three new classes have been introduced in version 1.2 of the MMS conformance document: *image rich, video basic* and *video rich*. For the definition of these content classes, codecs and formats have been categorized into 10 media types as shown in Table 5.4. This categorization was initially introduced in [3GPP-26.140] and is used in the subsequent sections of this book.

A multimedia message is conformant to a given content class if the following conditions are fulfilled:

- All media objects contained in the message are of a media format/codec allowed for the given class. In addition, media objects must fulfil the format requirements in terms of image resolutions, size, and so on.
- The message fulfils the requirements of the message class definition in terms of maximum message size allowed (30 kB, 100 kB or 300 kB) and applicability of the Digital Rights Management (DRM) mechanism (e.g. OMA DRM forward-lock cannot be applied to messages expected to conform to the requirements of the image basic class).

The five message content classes are defined in a hierarchical manner in which sophisticated classes are built up from basic classes by extending the set of supported

---

[2] MMS introduces two concepts of classes: message classes (advertisement, personal, informational, auto) and message content classes (as described in this section). They are two distinct concepts.

**Table 5.4**  Message content classes in the core message content domain

|                          | Class text          | Class image basic        | Class image rich         | Class video basic        | Class video rich         |
|--------------------------|---------------------|--------------------------|--------------------------|--------------------------|--------------------------|
| Text                     | US-ASCII, UTF-8, UTF-16 | US-ASCII, UTF-8, UTF-16 | US-ASCII, UTF-8, UTF-16 | US-ASCII, UTF-8, UTF-16 | US-ASCII, UTF-8, UTF-16 |
| Still image              | None                | Baseline JPEG (JFIF)     | Baseline JPEG (JFIF)     | Baseline JPEG (JFIF)     | Baseline JPEG (JFIF)     |
| Bitmap image             | None                | GIF87a, GIF89a, WBMP     | GIF87a, GIF89a, WBMP     | GIF87a, GIF89a, WBMP     | GIF87a, GIF89a, WBMP     |
| Speech[a]                | None                | AMR narrowband           | AMR narrowband           | AMR narrowband           | AMR narrowband           |
| (Music) Audio            | None                | None                     | None                     | None                     | None                     |
| Synthetic audio          | None                | None                     | SP-MIDI                  | SP-MIDI                  | SP-MIDI                  |
| Video[b]                 | None                | None                     | None                     | H.263 with AMR-NB (.3GP) | H.263 with AMR-NB (.3GP) |
| Vector graphics          | None                | None                     | None                     | None                     | None                     |
| Personal Information Manager | None            | vCard and vCalendar      | vCard and vCalendar      | vCard and vCalendar      | vCard and vCalendar      |
| Scene description        | MMS SMIL            | MMS SMIL                 | MMS SMIL                 | MMS SMIL                 | MMS SMIL                 |
| Support for OMA DRM – forward-lock | No        | No                       | Yes                      | Yes                      | Yes                      |
| Message size             | ≤30 kB              | ≤30 kB                   | ≤100 kB                  | ≤100 kB                  | ≤300 kB                  |
| Max image resolution     | Not applicable      | 160 × 120                | 640 × 480                | 640 × 480                | 640 × 480                |

[a] AMR narrowband is the recommended speech codec for MMS clients conforming to 3GPP requirements. For MMS clients conforming to Third Generation Partnership Project 2 (3GPP2) requirements, OMA recommends the use of the 13-K speech codec instead.

[b] An H.263 video clip with an optional AMR audio track is transported in a 3GP file as defined by the 3GPP. Alternatively, 3GPP2 compliant MMS clients use an alternative file format known as 3GP2.

formats/codecs and by allowing larger messages to be exchanged. Note that the hierarchical nature of message content classes may not be applicable to the definition of additional classes in the future.

None of the classes allows the support of formats/codecs for synthetic video and (music) audio. Future classes that are expected to be introduced in the next MMS conformance document will cover these aspects. Candidate formats/codecs for synthetic video and music audio are SVG (tiny profile) and AMR-wideband, respectively.

The MMSC can perform content adaptation to a multimedia message conformant to one of the classes, so it becomes conformant to another class better handled by the receiving device.

### 5.2.3 MMS Client Conformance to Message Content Classes

MMS clients can be categorized according to their support in terms of message creation, submission, retrieval and presentation (i.e. rendering) of messages compliant to the three message content classes.

- *Creation conformance*: An MMS client is said to be creation conformant towards a given class (text, image basic, image rich, video basic or video rich) if the client allows the insertion of any media formats/codecs allowed for the given message content class (see Table 5.4) and if, of course, messages created by the MMS client are conformant to the given message class. A creation conformant MMS client typically warns the user if the created message diverges from the requirements of the given content class.
- *Submission conformance*: An MMS client is said to be submission conformant towards a given class if the client is able to submit any multimedia message compliant to this given class. A submission conformant MMS client typically warns the user if the submitted message diverges from the requirements of the given content class.
- *Retrieval conformance*: An MMS client is said to be retrieval conformant towards a given class if the client is able to retrieve any multimedia message compliant to this given class.
- *Presentation conformance*: An MMS client is said to be presentation conformant towards a given class if the client is able to render all media objects contained in any multimedia message compliant to this given class.

An MMS client is said to be *fully conformant* to a given content class if it is creation, submission, retrieval and presentation conformant to the given content class. An MMS client is said to be *partially conformant* to a given content class if it is compliant in creation, submission, retrieval or presentation but not conformant to all four aspects. For instance, an observation camera, which is conformant to the image basic class in creation and submission only is said to be partially conformant to the image basic class.

MMS clients may support three distinct message creation modes: restricted, warning and free. With the *restricted* mode, the MMS client does not allow the creation of messages that do not belong to the message content class it is compliant to. With the *warning* mode, the MMS client allows the creation of messages that do not belong to the message content class it is compliant to. However, in this mode, a warning is given to the user upon creation of such messages. With the *free* mode, the creation of messages not belonging to the message class the MMS client is compliant to is allowed without warning given to the user. These modes may be pre-configured, configured by the user or configured by the operator via a device management mechanism (see Section 4.7.3).

## 5.3 Media Types, Formats and Codecs

This section presents the different media types supported in the context of MMS. MMS devices may support one or more media types (still image, video, speech, etc.). And for each supported media type, the MMS device supports at least one media format/codec (e.g. AMR for speech, H.263 for video, GIF and WBMP for bitmap images, etc.).

Appendix A provides a list of corresponding format/codec content types.

## 5.3.1 Text

Like other media objects, text in multimedia messages is contained in message body parts. Each text body part is characterized with a content type identifying one of the available character sets. The Internet Assigned Numbers Authority (IANA) publishes, in [IANA-MIBEnum], the list of character sets along with unique character set identifiers known as MIBEnum. Regardless of the message content class, [OMA-MMS-Conf] (version 1.2) mandates the support of the three character sets listed in Table 5.5.

MMS clients support the UTF-16 character set for backward compatibility with first commercial implementations. However, for interoperability reasons, it is recommended not to use UTF-16 for encoding text in multimedia messages.

## 5.3.2 Bitmap and Still Images

In the context of MMS, several formats can be used for representing images in multimedia messages. *Bitmap images* are expected to become widely used in MMS because of the large availability of such image formats over the Internet. Bitmap images are exact pixel-by-pixel mappings of the represented image (except if a lossy compression is used to reduce the bitmap size). Common formats for bitmap images are GIF, the Portable Network Graphic (PNG) format and the Wireless BitMaP (WBMP) format [WAP-237]. PNG is a file format whose definition is published in the form of a World Wide Web Consortium (W3C) recommendation [W3C-PNG]. PNG is used for representing bitmap images in a lossless fashion. PNG ensures a good compression ratio and supports image transparency. PNG represents a patent-free substitute for the GIF format. OMA mandates the support of GIF (GIF87a and GIF89a) and WBMP

**Table 5.5**  Character sets

| Character set | MIBEnum |
|---|---|
| US-ASCII | 3 |
| UTF-8 | 106 |
| UTF-16[a] | 1015 |

[a] [OMA-MMS-Conf] (version 2.0) indicates that MIBEnum for UTF-16 is 1000. This is an error since MIBEnum 1000 identifies UCS-2. This has been the source of interoperability problems for initial MMS solutions. This error is corrected in version 1.2 of the same document.

for bitmap images [OMA-MMS-Conf] (version 1.2) for all message content classes, except for the text class.

On the other hand, OMA mandates the support of JPEG with the JFIF file format for *still images* [OMA-MMS-Conf] (version 1.2) for all message content classes, except for the text class.

---

**BOX 5.2   Web resources for images**

PNG at W3C: http://www.w3c.org/Graphics/PNG/

---

For the composition of messages, the user has access to images stored locally in the handset (e.g. photo album). In addition, the MMS client often has access to photos taken with a digital camera (built into the phone or as an external accessory). Digital cameras allow the capture of photos according to various resolution modes. It is very common to refer to VGA[3] display modes when specifying the capture resolution of a camera. Image resolution modes shown in Table 5.6 are commonly supported by MMS-enabled phones (resolution is expressed in the number of horizontal pixels × the number of vertical pixels).

In addition to supported capture and display resolutions, an MMS phone is also characterized by the colour depth of its display screen and of its digital camera capture capabilities. Standardization organizations do not make any recommendation regarding the minimum colour depth to be supported by MMS devices. However, one can observe that the lowest supported colour depth for the display screen of available devices is 8 bits (256 colours). The majority of the first MMS wave phones support a minimum colour depth of 12 bits (4096 colours) and several more advanced MMS phones already support a colour depth of 16 bits (65,536 colours). Chapter 8 reviews the capabilities of MMS phones available today on the market.

### 5.3.3 Vector Graphics

In the future, multimedia messages will contain vector graphics. Vector graphics are based on descriptions of graphical elements composing the represented synthetic

**Table 5.6**   Image resolutions

| Display mode | Resolution |
| --- | --- |
| VGA | 640 × 480 |
| QVGA[a] | 320 × 240 |
| QQVGA | 160 × 120 |
| QQQVGA | 80 × 60 |

[a] QVGA (stands for Quarter VGA)

---

[3] Video Graphics Array (VGA) was introduced in 1987 by IBM and has become the accepted minimum standard for PC display systems.

image/animation. These descriptions are usually made using instructions similar to those of a programming language. Vector graphics instructions are processed by a graphics processor to reconstruct graphical elements contained in the represented image/animation. Metafile and the emerging SVG are two well-known vector graphic formats used for representing images. SVG is an open standard derived from eXtensible Markup Language (XML) and published as a W3C recommendation [W3C-SVG]. Advantages of SVG include the possibility of dynamically scaling the represented image according to the capabilities of the receiving device (e.g. screen size, frame rate). Furthermore, the size of SVG representing synthetic images/animations can be very low compared to the size of equivalent bitmap/still image and video representations.

However, scalable graphic formats are not appropriate for representing all types of images/animations. For instance, photographs and recorded video clips are usually not well represented with vector graphics. Representing photographs with a scalable vector graphic format may lead to very large representations, larger than equivalent bitmap/still image representations. Additional processing capabilities are usually required to render vector graphics (integer and floating point calculations). In the context of MMS, SVG looks to be the most suitable vector graphics format and 3GPP mandates the support for SVG tiny profile for devices supporting this media type (from Release 5). Furthermore, 3GPP recommends the support of SVG basic profile.

Note that, at the time of writing, none of the MMS-capable devices available on the market supported vector graphics for MMS and OMA had not yet defined a message content class with support of vector graphics [OMA-MMS-Conf] (version 1.2). The support of vector graphics in MMS is expected to be elaborated in the next version of [OMA-MMS-Conf].

---

**BOX 5.3   Web resources for vector graphics**

SVG at W3C: http://www.w3c.org/Graphics/SVG/

---

### 5.3.4 Speech

Several audio codecs are used in the context of MMS. Audio codecs are usually classified using three media types as defined below:

- *Speech audio* codecs used to represent speech samples such as voice memos.
- *Audio* codecs used to represent audio clips including recorded music teasers.
- *Synthetic audio* formats consist of specifying commands instructing a synthesizer on how to render sounds. Such formats are used for the representation of melodies.

For the *speech audio* media type, Adaptive Multirate (AMR) has long been the codec of choice. The AMR codec is typically used to represent speech in mobile networks for voice communications. This codec can adapt its rate according to available resources. AMR compresses linear-PCM speech input at a sample rate of 8 kHz adaptively to one of eight data rate modes: 4.75 kbps, 5.5 kbps, 5.9 kbps, 6.7 kbps, 7.4 kbps, 7.95 kbps, 10.2 kbps and 12.2 kbps. Note that AMR, when configured at given rates, becomes directly compatible with technical characteristics of other codecs specified by several standardization organizations. For instance, AMR, at the 12.2 kbps rate, is compatible with the Enhanced Full Rate (EFR) speech codec defined by the European Telecommunications Standard Institute (ETSI) for GSM. Furthermore, AMR, at the 7.4 kbps rate, becomes compatible with the IS-136 codec. Finally, AMR, configured at the 6.7 kbps rate, becomes compatible with the speech codec used in the Japanese PDC.

This initial specification of the AMR codec is also referred to as the AMR narrow-band (AMR-NB). AMR-NB is suitable for representing recorded speech and does not have the capabilities of representing adequately recorded music teasers. In the context of MMS, AMR data is stored and transported in multimedia messages according to the file format specified by IETF in [RFC-3267].

An extension of AMR-NB, known as AMR-wideband (AMR-WB), can be used for the representation of music clips, in addition to representing voice with a better quality than AMR-NB. AMR-WB compresses linear-PCM speech/music input at a sample rate of 16 kHz adaptively to a multitude of data rate modes from 6.6 to 23.85 kbps.

OMA mandates the support of AMR-NB for speech for all message content classes [OMA-MMS-Conf] (version 1.2), except for the text class.

### 5.3.5 Audio and Synthetic Audio

For the *audio* media type, MPEG-4 AAC low complexity codec is emerging as one appropriate codec. AMR-WB can also be used for the representation of music teasers. MP3 stands for MPEG Layer-3 and is an audio-compressed format. MP3 is based on perceptual coding techniques that address the perception of sound waves by the human ear. OMA has not yet defined a message content class with support of (music) audio [OMA-MMS-Conf] (version 1.2). The support of codecs for (music) audio in MMS is expected to be elaborated in the next version of [OMA-MMS-Conf].

For the *synthetic audio* media type, MIDI and iMelody formats have been used in the context of MMS. Since the release of MIDI 1.0 in 1982 [MMA-MIDI], MIDI has become the most widely used synthetic music standard among musicians and composers. The MIDI standard encompasses not only the specifications of the connector and cable for interconnecting MIDI-capable devices but also the format of messages exchanged between these devices. Only the format of MIDI messages is of interest in the context of MMS.

Melodies in the MIDI format are represented by a sequence of instructions that a sound synthesizer can interpret and execute. In an MMS device, this MIDI sound synthesizer may be implemented either as a software-based synthesizer or as a hardware MIDI chipset. Instructions rendered by the sound synthesizer are in the form of MIDI messages. For instance, a MIDI message can instruct the synthesizer to use a specific instrument or to play a given note. In theory, MIDI messages can be sent in real time to the synthesizer.

In the scope of MMS, the MIDI melody is first conveyed as part of a message and later rendered by the sound synthesizer when requested by the user. For this purpose, additional timing information needs to be associated with MIDI messages in order to tell the synthesizer when to play the melody notes. To achieve this, timing information along with MIDI messages are formatted as Standard MIDI Files (SMF). An SMF-formatted melody belongs to one of the three following SMF formats:

- *SMF format 0*: With this format, all MIDI messages are stored in one single track.
- *SMF format 1*: With this format, MIDI messages are stored as a collection of tracks.
- *SMF format 2*: This format is seldom supported by sound synthesizers.

Available MIDI synthesizers have various rendering capabilities. Some synthesizers only support a low level of polyphony[4] while others support a high level of polyphony. With MMS, the synthesizer is embedded in a mobile device, which usually has limited processing capabilities. In most cases, device resources available for the MIDI synthesizer to render sounds depend on the state of other applications being executed on the device. To cope with devices having various capability levels, the MIDI format has evolved to support the concept of scalable polyphony. This evolved MIDI format is known as the Scalable Polyphony MIDI (SP-MIDI) [MMA-SP-MIDI]. Scalable polyphony consists of indicating, in the melody, what instructions can be dropped without significant quality degradation when the receiving device is running short of resources. For this purpose, the scalable polyphony information is provided as part of a specific system-exclusive message known as a Maximum Instantaneous Polyphony (MIP) message. The scalable polyphony information indicates the note usage and priority for each MIDI channel. For instance, a melody composer can generate a MIDI melody that can be best rendered with a maximum polyphony of 16 but can still be rendered by synthesizers supporting a maximum polyphony of 10 or 8 by dropping low-priority instructions. Regarding the use of Scalable Polyphony MIDI (SP-MIDI), the MIDI Manufacturers Association (MMA), in cooperation with the Association of Music Electronics Industries (Japan), has defined an SP-MIDI profile to be supported by devices with limited capabilities (e.g. mobile devices). This profile, known as the 3GPP profile for SP-MIDI [MMA-SP-MIDI], provides a means of ensuring interoperability between devices supporting a level of note-polyphony ranging from 5 to 24. This profile identifies MIDI messages and SP-MIDI features, which shall be supported by mobile devices in order to ensure satisfactory interoperability.

---

[4] The polyphony of a sound synthesizer refers to its ability to render several notes or voices at a time. Voices can be regarded as units of resources required by the synthesizer to produce sounds according to received MIDI instructions. Complex notes for given instruments typically require more than one synthesizer voice to be rendered.

Note that the iMelody format is not a format suggested for MMS by standardization organizations. Nevertheless, several MMS-capable devices available on the market do support iMelody tunes in multimedia messages (see Chapter 8).

OMA mandates the support of SP-MIDI for synthetic audio for image rich, video basic and video rich content classes [OMA-MMS-Conf] (version 1.2).

### 5.3.6 Video

First MMS devices did not support video clips in multimedia messages. At present, several MMS devices do have support for video and this trend is expected to grow. OMA mandates the support of [ITU-H.263] for the video basic and video rich content classes [OMA-MMS-Conf] (version 1.2). MPEG-4 represents an alternative candidate for the future (see Box 5.4).

The International Telecommunication Union (ITU) has published several video standards including H.261 (late 1980/early 1990s), H.263 (mid-1990s) and more recently H.26L. In the context of MMS, messages conformant to the video basic or video rich class may contain video clips encoded according to H.263 profile 0 level 10. The maximum size of video clips is limited by the maximum message size inherent to each message content class: 100 kB for video basic and 300 kB for video rich.

An MMS phone supporting video is typically capable of displaying a digital video in one of the video frame resolutions listed in Table 5.7. A mobile device supporting H.263 profile 0 level 10 is able to support at least the Sub-QCIF and the QCIF video resolutions.

In the context of MMS, standardization organizations recommend the use of a specific file format for timed multimedia (e.g. video with audio) known as the 3GPP file format (3GP) and defined in [3GPP-26.234].

### 5.3.7 Personal Information Manager Objects

If a Personal Information Manager (PIM) is present in the MMS mobile device, then electronic business cards and calendaring/scheduling information may be exchanged with MMS. PIM objects are represented with vCard and vCalendar formats.

**Table 5.7**  Video frame resolutions

| Display mode | Resolution |
| --- | --- |
| 4CIF | 704 × 546 |
| CIF | 352 × 288 |
| QCIF[a] | 176 × 144 |
| Sub-QCIF | 128 × 96 |

[a] QCIF (stands for Quarter CIF)

---

**BOX 5.4   MPEG video codecs**

MPEG stands for Moving Pictures Expert Group and is an organization that develops technical specifications for representing and transporting video. The first specification from this organization was MPEG-1, published in 1992. MPEG-1 allows video players to render video in streaming mode. MPEG-2, introduced in 1995, supersedes MPEG-1 features and is mainly used for compression and transmission of digital television signals. In December 1999, the group released the specification for MPEG-4 (ISO/IEC 14496), based on an object-oriented paradigm, where objects are organized in order to compose a synthetic scene. This is a major evolution from previous MPEG formats. The compact MPEG-4 language used for describing and dynamically changing scenes is known as the Binary Format for Scenes (BIFS). BIFS commands instruct the MPEG-4 player to add, remove and dynamically change scene objects in order to form a complete video presentation. Such a technique allows the representation of video scenes in a very compact manner. MPEG-4 also supports streaming in low bit rate environments. MPEG-4 has proved to provide acceptable streaming services over 10 kbps channels. Mobile networks often provide very variable and unpredictable levels of resources for services. To cope with these network characteristics, MPEG-4 can prioritize objects to transmit the most important objects only when the system is running short of resources. Owing to the limitations of available mobile devices, 3GPP recommends the use of the visual simple profile level 0 of MPEG-4: the simplest of available profiles and levels for MPEG-4. OMA has not yet defined a message content class with support of MPEG-4 [OMA-MMS-Conf] (version 1.2).

**Web Resources for MPEG**

Moving Pictures Expert Group: http://mpeg.telecomitalialab.com
MPEG-4 Industry Forum: http://www.m4if.org

---

The *vCard* format is used for representing electronic business cards [IMC-vCard]. This format is already widely used with Personal Digital Assistants (PDAs) and is becoming the de facto format for the exchange of electronic business cards over infrared links. It is also becoming common to attach a vCard object, as a signature, to an email message. A vCard object contains basic contact details such as last name, first name, postal and electronic addresses, phone, mobile and fax numbers, and so on. It may also contain more sophisticated elements such as photographs or company logos.

On the other hand, the *vCalendar* format is used to represent items generated by calendaring and scheduling applications [IMC-vCalendar]. As for the vCard format, it is widely used with PDAs and is becoming the de facto format for the exchange of calendaring/scheduling information. A vCalendar object is composed of one or more elements of types *event* and *todo*. An *event* is a calendaring/scheduling element representing an item in a calendar. A *todo* is a calendaring/scheduling element representing an action item or assignment.

## 5.4 Scene Description

Previous sections have described how several media objects can be included in a multimedia message. In order to enrich the user experience, it is common to organize the way media objects are rendered on the receiving device screen and when they should be rendered over a common timeline. This allows the creation of truly multimedia presentations in which media objects are choreographed in a meaningful manner on the receiving device. This media object organization, also called *scene description*, is defined using a format/language such as the Synchronized Multimedia Integration Language (SMIL) or XHTML. The minimal subset of SMIL defined in [OMA-MMS-Conf], also known as *MMS SMIL*, became the de facto scene description representation for available MMS devices and has now gained the status of OMA standard. In the meantime, 3GPP has elaborated a more sophisticated SMIL profile for MMS. This profile is expected to become the standardized profile for future devices. At the time of writing, none of the available MMS devices supported the 3GPP SMIL profile for MMS.

### 5.4.1 Introduction to SMIL

The Synchronized Multimedia Integration Language (SMIL), pronounced 'smile', is an XML-based language published by W3C. A major version of this language, SMIL 2.0 [W3C-SMIL], is organized around a set of modules defining the semantics and syntax of multimedia presentations (for instance, modules are available for the timing and synchronization, layout and animation, etc.). SMIL is not a codec nor a media format but rather a technology allowing media integration. With SMIL, the rendering of a set of media objects can be synchronized over time and organized dynamically over a predefined graphical layout to form a complete multimedia presentation. SMIL is already supported by a number of commercial tools available for personal computers including RealPlayer G2, Quicktime 4.1 and Internet Explorer (from version 5.5).

Owing to small device limitations, a subset of SMIL 2.0 features has been identified by W3C to be supported by devices such as PDAs. This subset, called SMIL *basic profile*, allows small appliances to implement some of the most useful SMIL features without having to support the whole set of SMIL 2.0 instructions. Unfortunately, the SMIL basic profile appeared to be still difficult to implement on MMS mobile devices. To cope with this difficulty, a group of manufacturers designed an even more limited SMIL profile, known as the MMS SMIL, supported by most MMS phones that support scene descriptions. The Open Mobile Alliance is now the organization in charge of maintaining and publishing the MMS SMIL specification as part of [OMA-MMS-Conf] (available from MMS 1.1).

In the meantime, 3GPP has produced specifications for a *3GPP SMIL* profile, also known as the packet-switched streaming SMIL profile (PSS SMIL profile). The 3GPP SMIL profile is to become the future standard profile for all MMS-capable devices. The MMS SMIL is an interim profile until devices can efficiently support the 3GPP SMIL profile. The 3GPP SMIL profile is still a subset of SMIL 2.0 features, but a superset of the SMIL basic profile, and is published in [3GPP-26.234].

Designers of SMIL multimedia presentations can

- describe the temporal behaviour of the presentation
- describe the layout of the presentation on a screen
- associate hyperlinks with media objects
- define conditional content inclusion/exclusion on the basis of system/network properties.

---

**BOX 5.5    Resources for SMIL**

A comprehensive two-part tutorial on SMIL is identified in the further reading section of this chapter (Bulterman, 2001, 2002).
SMIL at W3C: http://www.w3c.org/AudioVideo

---

### 5.4.2 Organization of SMIL 2.0

A major version of the language, SMIL version 2.0, has been publicly released by W3C in August 2001. The 500-page SMIL 2.0 specifications define a collection of XML tags and attributes that are used to describe temporal and spatial coordination of one or more media objects that form a multimedia presentation. This collection is organized into 10 major functional groups as shown in Figure 5.3.

Each functional group is composed of several modules (from 2 to 20). The aim of this SMIL organization is to ease the integration of SMIL features into other XML-derived languages. A number of profiles have been defined on the basis of this organization. An SMIL profile is a collection of modules. So far, several profiles have been introduced such as the SMIL 2.0 language profile, XHTML + SMIL profile and the SMIL 2.0 basic profile (as introduced earlier).

### 5.4.3 Spatial Description with SMIL

SMIL 2.0 content designers are able to define sophisticated spatial layouts. The presentation rendering space is organized by regions. Each region in the layout can accommodate a graphical object such as an image, a video clip or some text. Regions

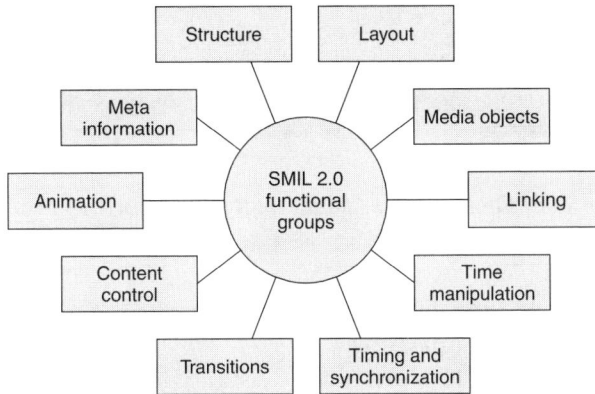

**Figure 5.3**  SMIL 2.0 functional groups

```
<layout>
        <root-layout width="128" height="128"/>
        <region id="Image" width="128" height="72" left="0" top="0"/>
        <region id="Text" width="128" height="56" left="0" top="72"/>
</layout>
```

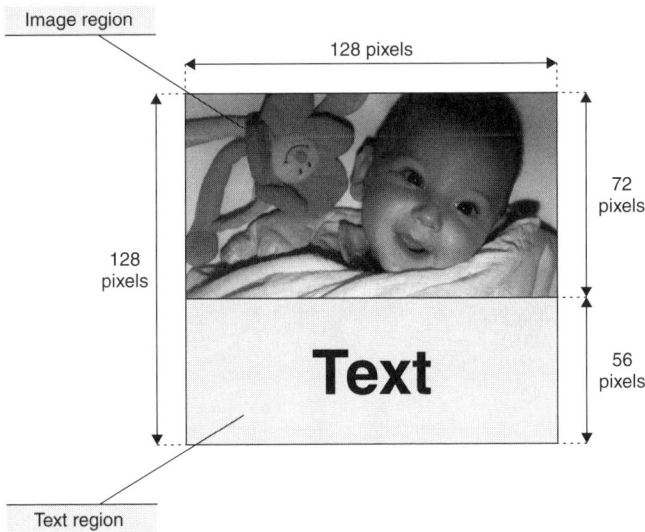

**Figure 5.4**  SMIL/layout container. Smiling Louise before bedtime

can be nested in each other in order to define advanced multimedia presentations.
The tag root-layout defines the main region of the presentation. Sub-regions to
be nested within the main region are defined with the region tag. The SMIL example
in Figure 5.4 shows how two sub-regions, one accommodating an image and the other
one some text, can be defined within the main region.

### 5.4.4 Temporal Description with SMIL

Objects in a SMIL presentation can be synchronized over a common timeline. For this purpose, a set of time containers can be used such as the sequential and the parallel time containers:

- *Sequential time container*: This container identified by the `seq` tag enables the sequencing of an ordered list of objects. Each object is rendered in turn and the rendering of each object starts when the rendering of the previous object has terminated. An absolute time duration for the rendering of each object may also be specified as shown in Figure 5.5.
- *Parallel time container*: This container, identified by the `par` tag, enables the rendering of several objects in parallel as shown in Figure 5.6.

In a scene description, containers can be nested in order to create a whole hierarchy of time containers for the definition of sophisticated multimedia presentations.

### 5.4.5 SMIL Basic Profile

As indicated in previous sections, W3C has defined a SMIL basic profile for SMIL 2.0. The SMIL basic profile is a subset of the full set of SMIL 2.0 features, appropriate

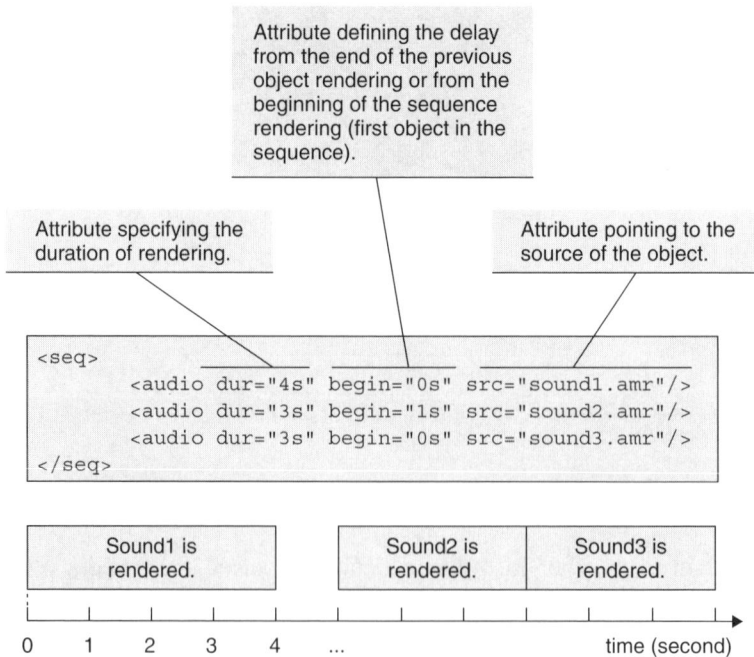

**Figure 5.5** SMIL/sequential time container

```
<par>
        <audio dur="4s" begin="0s" src="sound1.amr"/>
        <audio dur="3s" begin="1s" src="sound2.amr"/>
        <audio dur="3s" begin="0s" src="sound3.amr"/>
</par>
```

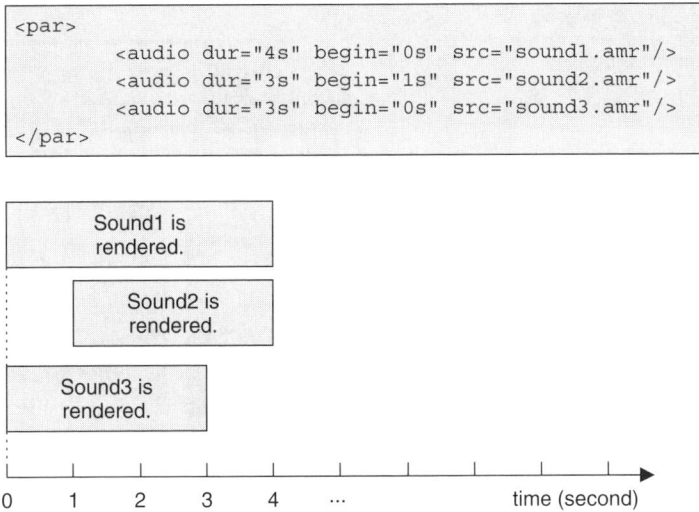

**Figure 5.6**   SMIL/parallel time container

for small appliances such as PDAs. In the short term, this profile does not appear to be suitable for mobile devices, which still have very limited capabilities. In order to cope with such limitations, the MMS SMIL defined in the next section is used instead.

### 5.4.6 MMS SMIL and the OMA Conformance Document

With early MMS standards, manufacturers lacked an appropriate message scene description language: a language rich enough to allow the design of basic multimedia presentations but simple enough to be supported by devices with very limited capabilities. To fulfil this requirement, MMS-IOP, an informal working group of industrial partners, designed a profile for SMIL that fulfils the need of first MMS-capable devices. The profile is known as the MMS SMIL and was initially defined outside standardization processes in a document known as the MMS conformance document, as introduced earlier in Section 5.2. In addition to the definition of MMS SMIL, the conformance document also provides a set of recommendations to ensure interoperability between MMS-capable devices produced by different manufacturers.

With MMS SMIL, a multimedia presentation is composed of a sequence of slides (a slideshow) as shown in Figure 5.7. All slides from the slideshow have the same region configuration. A region configuration can contain at most one text region (named Text), at most one image/video region (named Image) and each slide can be associated with at most one audio clip (speech or synthetic audio). The audio clip may be encapsulated into the video clip (e.g. 3GP file) or inserted in the message as an independent media object (e.g. AMR or SP-MIDI file). However, [OMA-MMS-Conf] (version 1.2) forbids the inclusion of both an independent audio clip and a video clip in the same slide, even if the video clip does not contain any audio track.

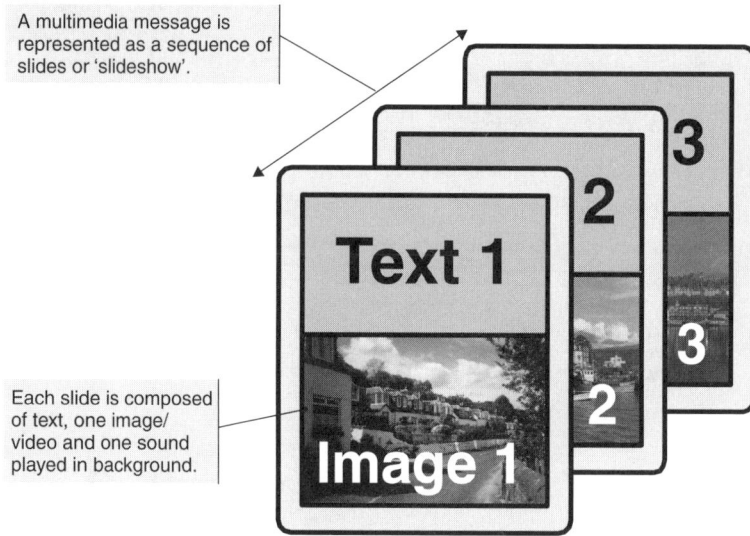

A multimedia message is represented as a sequence of slides or 'slideshow'.

Each slide is composed of text, one image/ video and one sound played in background.

**Figure 5.7**   Message presentation with MMS SMIL

For slideshows containing both, a text region and an image region, the term *layout* refers to the way regions are spatially organized: vertically (portrait) or horizontally (landscape). Mobile devices have various screen sizes and it may therefore happen that a particular layout for a message scene description does not fit in the device display. In this situation, the layout may be overridden by the receiving device (from portrait to landscape and vice versa).

A message presentation may contain timing settings that allow an automatic slideshow rendering, during which the switching from one slide to the following is performed automatically after a time period defined as part of the message scene description. However, MMS phones often allow an interactive control by ignoring the timing settings and by allowing the user to switch from one slide to the following by simply pressing a key.

The MMS conformance document identifies SMIL 2.0 features that can be used for constructing the slideshow. A slideshow should have a sequence of one or more parallel time containers (instructions within the two tags `<par>...</par>`). Each parallel time container represents the definition of one slide. Each media object identified inside the parallel time container represents one component of a slide. Supported language elements and attributes in MMS-SMIL are listed in Table 5.8.

In addition to the definition of MMS SMIL, the MMS conformance document also contains the following rules:

• Dimensions of regions inside the root region should be expressed in absolute terms (i.e. pixels) or in percentage relative to the dimensions of the root region. It is not recommended to mix absolute and relative dimensions in the same scene description.

**Table 5.8** Language elements/attributes in MMS SMIL

| Functional groups | Elements | Attributes | Content model |
|---|---|---|---|
| | Layout | | region, layout |
| Layout | Region | left, top, height, width, fit, id | |
| | Root-layout | width, height | |
| | Text | src, region, alt, begin, end | |
| | Img | src, region, alt, begin, end | |
| Media | Audio | src, alt, begin, end | |
| | Video | src, region, alt, begin, end | |
| | Ref | src, region, alt, begin, end | |
| | Smil | | head, body |
| Structure | Head | | layout |
| | Body | | par |
| Timing and Synchronization | Par | dur | text, img, audio, ref |
| Meta information | Meta | name, content | |

- The `src` attribute must refer to a media type compliant with the associated element. For instance, the value associated with the `src` attribute of an `img` element must refer to an image, and nothing else.
- Timing information should be expressed in integer milliseconds.
- Maximum image dimensions, for which interoperability is ensured, are $160 \times 120$ pixels for messages compliant to the image basic class and $640 \times 480$ for other content classes.

The MMS conformance document specifies that a slide may be composed of a text region only, of an image region only or of two regions, one for the text named `Text` and another one named `Image`. The region named `Image` can accommodate either an image or a video clip. This means that a slide cannot contain both an image and a video clip.

Considering the two types of layouts for messages containing both an image/video region and a text region, a slideshow can be formatted according to the six configurations shown in Figure 5.8. In its simplest form, a slideshow is configured for representing only an image/video or only text on each slide. Such configurations are shown in Figure 5.8 (a) and (b). With the portrait layout, the image/video region may be positioned at the bottom or at the top of the screen. Examples of portrait layouts are depicted in Figure 5.8 (c) and (d). With the landscape layout, the image/video

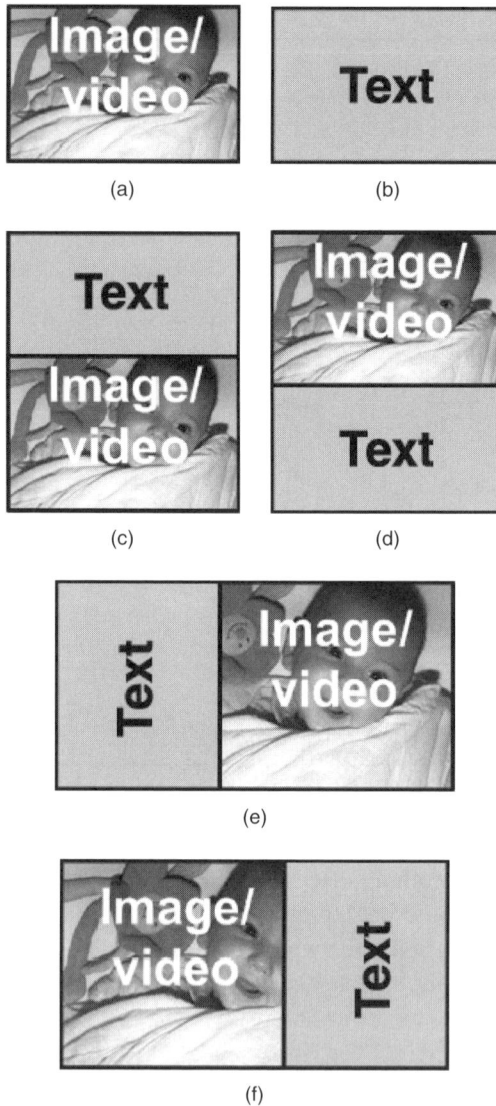

**Figure 5.8**  Scene description layouts

may be positioned at the right or at the left of the screen. Examples of landscape layouts are depicted in Figure 5.8 (e) and (f).

Figure 5.9 shows an example of MMS SMIL scene description corresponding to a two-slide configuration. The two slides are structured according to the portrait layout with the image/video region at the top and the text region at the bottom of the screen. In this example, the first slide contains an image (Image1.jpg), some text (Text1.txt) and an audio clip (sound1.amr), whereas the second slide contains a video clip (Video2.3gp) and some text (Text2.txt). The first slide is

The **SMIL presentation** definition is incorporated between the tags `<smil>` and `</smil>`. An SMIL presentation is composed of a header and a body.

The **presentation header** is delimited by the tags `<head>` and `</head>`. The header contains information such as the presentation title, the author name and the presentation layout.

The **presentation layout** is delimited by the tags `<layout>` and `</layout>`. In this example, the layout is composed of an upper region containing an image or a video clip and a lower region containing some text.

```
<smil>
    <head>

        <meta name="title" content="mms"/>
        <meta name="author" content="Gwenael Le Bodic"/>

        <layout>
                <root-layout width="128" height="128"/>
                <region id="Image" width="128" height="72" left="0" top="0"/>
                <region id="Text" width="128" height="56"  left="0" top="72"/>
        </layout>

    </head>

    <body>

        <par dur="4000ms"><!-- First message slide -->
                <img src="cid:Image1.jpg" region="Image"/>
                <text src="cid:Text1.txt" region="Text"/>
                <audio src="cid:Sound1.amr" />
        </par>

        <par dur="10000ms"><!-- Second message slide -->
                <video src ="cid:Video2.3gp" region="Image"/>
                <text src ="cid:Text2.txt" region="Text"/>
        </par>

    </body>

</smil>
```

The **presentation body** is delimited by the tags `<body>` and `</body>`. The body contains the definition of each slide.

The **message slide** defines the content of the message slide and the temporal synchronization between elements in the slide. The layout of the slide is defined in the message layout.

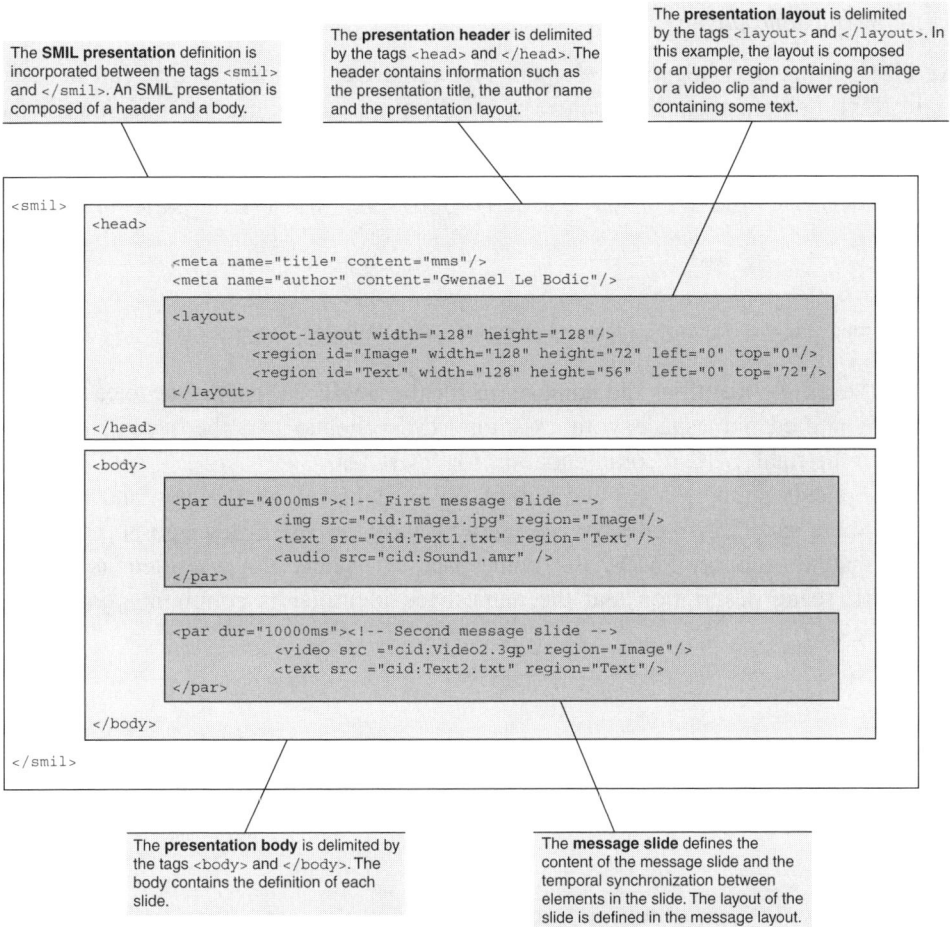

**Figure 5.9** Example of SMIL scene description

displayed during 4 seconds followed by the second slide for 10 seconds. The entire slide presentation time is therefore 14 seconds.

According to the MMS conformance document, if the multimedia message contains a scene description, then the `Content-type` header field of the multipart message header is set to `application/vnd.wap.multipart.related`; otherwise it is set to `application/vnd.wap.multipart.mixed`. Furthermore, if a presentation is available, the `start` parameter of the `Content-type` header field refers to the content identifier (`Content-ID` parameter) of the scene description present in the message and the `type` parameter indicates the format of the scene description (e.g. SMIL presentation) as shown below:

```
Content-type: application/vnd.wap.multipart.related;
                start = <0000>;
            type = 'application/smil'
```

### 5.4.7 SMIL Namespace

An XML namespace identifies element types and attribute names that can be used in an XML document. In the context of MMS, a namespace can be specified for the SMIL scene description as the `xmlns` attribute of the `smil` element as shown below:

```
<smil xmlns = "http://www.w3.org/2000/SMIL20/CR/Language">
```

or

```
<smil xmlns = "http://www.w3.org/2001/SMIL20/Language">
```

The first example identifies the namespace for the SMIL 2.0 candidate release (commonly identified namespace with existing MMS clients) and the second example identifies the final W3C recommendation for SMIL 2.0.

MMS clients support a limited subset of available element types and attribute names. Some early MMS clients even ignore totally the SMIL description. For MMS clients that do support SMIL, the namespace is usually not specified as part of the SMIL scene description and the namespace identifier is commonly omitted as shown below:

```
<smil>
```

If the namespace attribute is omitted from a scene description, then an SMIL parser assumes that the scene description element types and attribute names are the ones defined in SMIL 1.0.

### 5.4.8 Linking the Scene Description with Body Parts

Like basic media objects, a scene description is encapsulated in a message body part. The scene description refers to message media objects with a unique object identifier in the message and instructs how the referred media object should be rendered by the MMS client. The two most common object identifiers that are used are the `Content-ID` and the `Content-Location`.

With MMS-SMIL, a scene description is structured as a sequence of slides. Each slide definition, in the scene description, refers to message media objects displayed inline on the slide (for images, text and video clips) or rendered when the slide is displayed on the screen (for audio clips). Message media objects that are not referred to by the scene description are considered as message attachments. Message attachments are usually not displayed inline in message slides but the MMS client often provides features for storing attachments in the device local memory (e.g. internal flash memory or external flash card). This feature is particularly useful for exchanging PIM elements such as vCard or vCalendar objects.

### 5.4.8.1 Linking with Content Identifier

The most straightforward method for referring to body parts in a scene description consists of using the content identifier (`Content-ID` parameter) as a unique identifier for a message body part. As shown in Section 5.1.2, each body part is associated with a content identifier in the following form:

```
Content-ID: <body_part_006>
```

In the example above, the value used as content identifier is `body_part_006` (square brackets are removed). In an SMIL scene description, a reference to a given body part using the content identifier is made according to the following instruction:

```
<img src = "cid:body_part_006" region = "Image">
```

The prefix 'cid:' indicates that the method used for referring to the body part is the method based on content identifiers. In this example, the SMIL instruction '<img ... region = "Image">' indicates that the media object is an image to be inserted in the region named `Image`. Similarly, the SMIL instruction `<video ... region = "Image">` indicates that a video clip is inserted in the region named `Image`, whereas the instruction `<text ... region = "Text">` indicates that a text file is to be inserted in the region named `Text`. The instruction `<audio ... >` for including an audio clip in a slide makes no reference to a specific region.

Figure 5.9 shows a scene description whose references are made of content identifiers.

### 5.4.8.2 Linking with Content Location

Alternatively, the scene description can also refer to body parts using content locations. Each body part can be associated to a content location [RFC-2557] as shown below:

```
Content-location: Image1.jpg
```

In the example above, the value used as content location is `Image1.jpg`. In an SMIL scene description, a reference to a given body part using the content-location parameter is made according to the following instruction:

```
<img src = "Image1.jpg" region = "Image">
```

The fact that the prefix 'cid:' is missing from the reference means that the scene description uses content locations for referring to body parts. As for the content identifier method, the SMIL instruction '<img ... region = "Image">' indicates in which region of the slide the referred media object should appear.

Note that the value assigned to the content location parameter is also used by the message designer (or by the MMS client) to provide a default file name to the message media object in case the recipient wishes to store the object in the receiving device local memory (see Box 5.6).

Figure 5.10 below shows a complete scene description whose references are made of content locations.

### 5.4.9 Support of Video Streaming

It has been shown in this chapter how a media object (e.g. video clip) can be inserted in a multimedia message and linked to a particular region of a SMIL scene description. An alternative method consists of indicating in the multimedia message a reference to a media object located in a remote media server. This method is typically

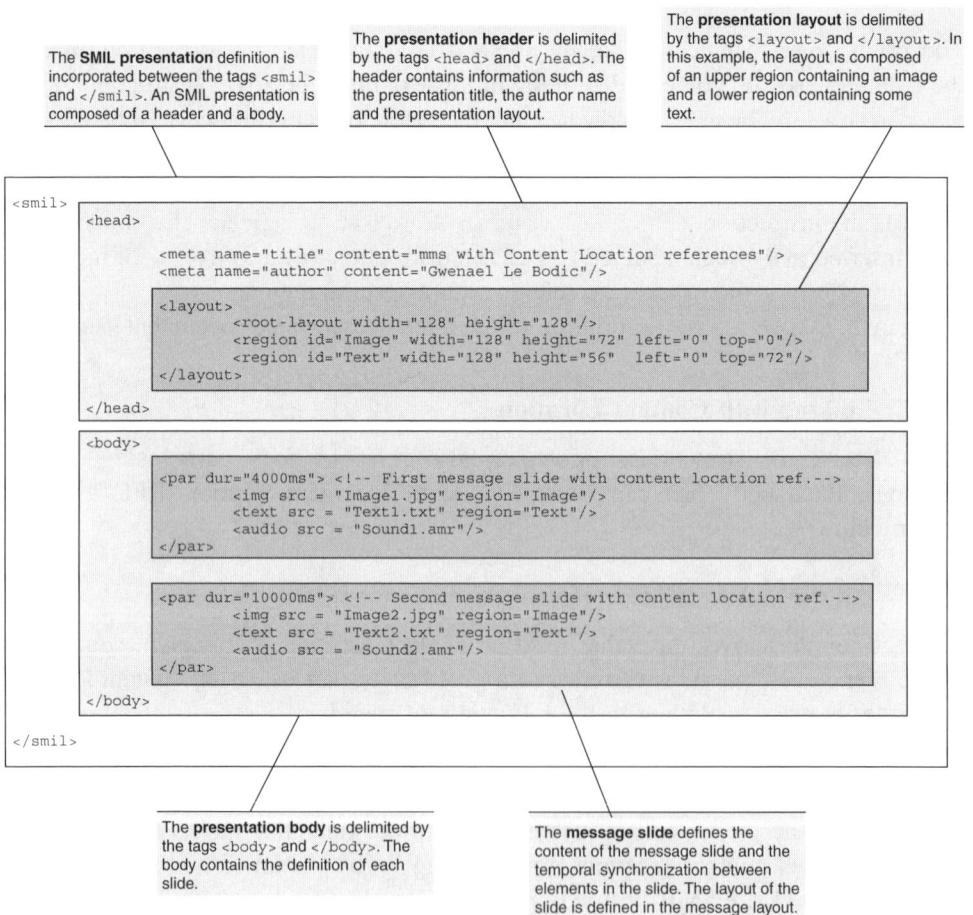

The **SMIL presentation** definition is incorporated between the tags `<smil>` and `</smil>`. An SMIL presentation is composed of a header and a body.

The **presentation header** is delimited by the tags `<head>` and `</head>`. The header contains information such as the presentation title, the author name and the presentation layout.

The **presentation layout** is delimited by the tags `<layout>` and `</layout>`. In this example, the layout is composed of an upper region containing an image and a lower region containing some text.

```
<smil>
    <head>

            <meta name="title" content="mms with Content Location references"/>
            <meta name="author" content="Gwenael Le Bodic"/>

            <layout>
                    <root-layout width="128" height="128"/>
                    <region id="Image" width="128" height="72" left="0" top="0"/>
                    <region id="Text" width="128" height="56"  left="0" top="72"/>
            </layout>

    </head>

    <body>

            <par dur="4000ms"> <!-- First message slide with content location ref.-->
                    <img src = "Image1.jpg" region="Image"/>
                    <text src = "Text1.txt" region="Text"/>
                    <audio src = "Sound1.amr"/>
            </par>


            <par dur="10000ms"> <!-- Second message slide with content location ref.-->
                    <img src = "Image2.jpg" region="Image"/>
                    <text src = "Text2.txt" region="Text"/>
                    <audio src = "Sound2.amr"/>
            </par>

    </body>

</smil>
```

The **presentation body** is delimited by the tags `<body>` and `</body>`. The body contains the definition of each slide.

The **message slide** defines the content of the message slide and the temporal synchronization between elements in the slide. The layout of the slide is defined in the message layout.

**Figure 5.10**   Scene description with content-location referencing

---

**BOX 5.6   Recommendations for linking body parts and the scene description**

With first MMS developments, referencing methods based on Content-ID and Content-Location were often misused. In many cases, the cid prefix was missing from the references made of content identifiers. To cope with this issue, assigning the same value to the content identifier and content location reduces the probability to face interoperability problems with receiving MMS clients. However, this is not always possible since the content location is also used for providing a default file name in case the recipient wishes to store the associated media object in the receiving device local memory (e.g. photo album).

Mixing the two methods by using content location and content identifiers for referring to body parts, in the same scene description, is possible but not recommended.

---

used for referring to a video clip to be streamed from a media server down to the mobile device during message viewing and so reducing significantly the message size. Section 4.12 explained that the reference to the media object is provided as part of a presentation description (in the SDP format) or as a direct reference to the media object. In the latter case, the direct reference can be indicated as part of the SMIL scene description. The scene description below shows part of a scene description instructing the MMS client to retrieve a video clip from a media server in streaming mode using RTSP and to render this video clip in the region named Image:

```
<video src = "rtsp://www.lebodic.net/video/sample.3gp" region =
   "Image">
```

The OMA conformance document [OMA-MMS-Conf] (version 1.2) does not cover streaming aspects for MMS. This should be elaborated in future versions of the document.

### 5.4.10 Support of Colour with SMIL

MMS SMIL does not provide support for slide colours (background colour and text foreground colour). However, SMIL 2.0 includes a set of element attributes for the support of slide colours. These attributes are optional in the scene description. A single slide background colour can be specified for the entire message by assigning a colour value to the background attribute of the root-layout tag as shown below:

```
<root-layout background-color = "yellow" width = "352" height =
   "144"/>
```

In addition, a text foreground colour may be specified for each slide text object. This is performed by assigning a colour attribute when linking the text object with the text region as shown below:

```
<text src = "Text1.txt" region = "Text">
      <param name = "foreground-color" value = "blue"/>
</text>
```

It is common to use the following 16 colour keywords defined in [W3C-HTML4]: aqua, black, blue, fuchsia, gray, green, lime, maroon, navy, olive, purple, red, silver, teal, white and yellow. In addition, RGB hexadecimal colour codes [W3C-sRGB] may also be used (values are prefixed with the hash mark '#') as shown below:

```
black  =    "#000000"          green  =    "#008000"
silver =    "#C0C0C0"          lime   =    "#00FF00"
gray   =    "#808080"          olive  =    "#808000"
white  =    "#FFFFFF"          yellow =    "#FFFF00"
maroon =    "#800000"          navy   =    "#000080"
red    =    "#FF0000"          blue   =    "#0000FF"
purple =    "#800080"          teal   =    "#008080"
fuchsia =   "#FF00FF"          aqua   =    "#00FFFF"
```

The support of colour is not specified in the MMS conformance document. Consequently, the support of text colours is not widely supported by MMS clients.

### 5.4.11 XHTML as an Alternative to SMIL

As an alternative to SMIL, eXtensible HTML (XHTML) is a language that can also be used for representing message scene descriptions. In particular, XHTML Mobile Profile (XHTML MP) [WAP-277] extends HTML Basic Profile published by W3C [W3C-XHTML-Basic]. HTML MP is a subset of HTML 1.1 but a superset of HTML Basic Profile. XHTML MP has been specifically tailored for resource constrained devices. However, HTML MP remains a suitable language for the definition of rich MMS scene descriptions.

The OMA conformance document [OMA-MMS-Conf] (version 1.2) does not identify XHTML as an alternative to SMIL for MMS clients. Consequently, existing MMS devices seldom provide support for XHTML as a scene description language for MMS.

## 5.5 Example of a Multimedia Message

As shown previously, the multipart structure of a multimedia message is represented in a binary form in order to be efficiently transported between the MMS client and the MMSC. This binary representation is directly derived from the MIME concepts introduced in Section 5.1.1. Figure 5.11 shows the textual representation of a multimedia

**MMS PDU header**
This header contains parameters such as message recipient(s), subject, MMS version, etc.

```
X-Mms-Message-Type: M-send.req
X-Mms-Transaction-ID: 0123456789
X-Mms-MMS-Version: 1.0
From:+33144556677/TYPE=PLMN
To:+33111223344/TYPE=PLMN
Subject:A stay in Velen.
Content-type: application/vnd.wap.multipart.related
                          start = <0000>
                          type  = "application/smil"
                          boundary=boundary1
```

**MMS PDU body**
The body contains the message contents including the SMIL scene description and related objects (text, sounds, images, etc.).

```
--boundary1
Content-type: application/smil
Content-ID: <0000>

<smil>
 <head>
  <layout>
   <root-layout width="160" height="140"/>
   <region id="Image" width="160" height="120" left="0" top="0"/>
   <region id="Text" width="160" height="20" left="0" top="120"/>
  </layout>
 </head>

 <body>
  <par dur="5000ms">
   <img src="cid:VelenS1.jpg" region="Image"/>
   <text src="cid:VelenS1.txt" region="Text"/>
   <audio src="cid:VelenS1.amr"/>
  </par>
  <par dur="10000ms">
   <img src="cid:VelenS2.gif" region="Image"/>
   <text src="cid:VelenS2.txt" region="Text"/>
   <audio src="cid:VelenS2.amr"/>
  </par>
 </body>
</smil>
```

*SMIL scene description*

```
--boundary1
Content-type: image/jpeg
Content-ID: <VelenS1.jpg>
… The binary representation of the image for the 1st slide …
```

```
--boundary1
Content-type: audio/amr
Content-ID: <VelenS1.amr>
… The binary representation of the audio clip for the 1st slide …
```

```
--boundary1
Content-type: text/plain
Content-ID: <VelenS1.txt>

Hi, I spent a week in Velen.
```

*1st slide contents*

```
--boundary1
Content-type: image/jpeg
Content-ID: <VelenS2.jpg>
… The binary representation of the image for the 2nd slide …
```

```
--boundary1
Content-type: audio/amr
Content-ID: <VelenS2.amr>
… The binary representation of the audio clip for the 2nd slide …
```

```
--boundary1
Content-type: text/plain
Content-ID: <VelenS2.txt>

A pleasant stay in a castle.
--boundary1--
```

*2nd slide contents*

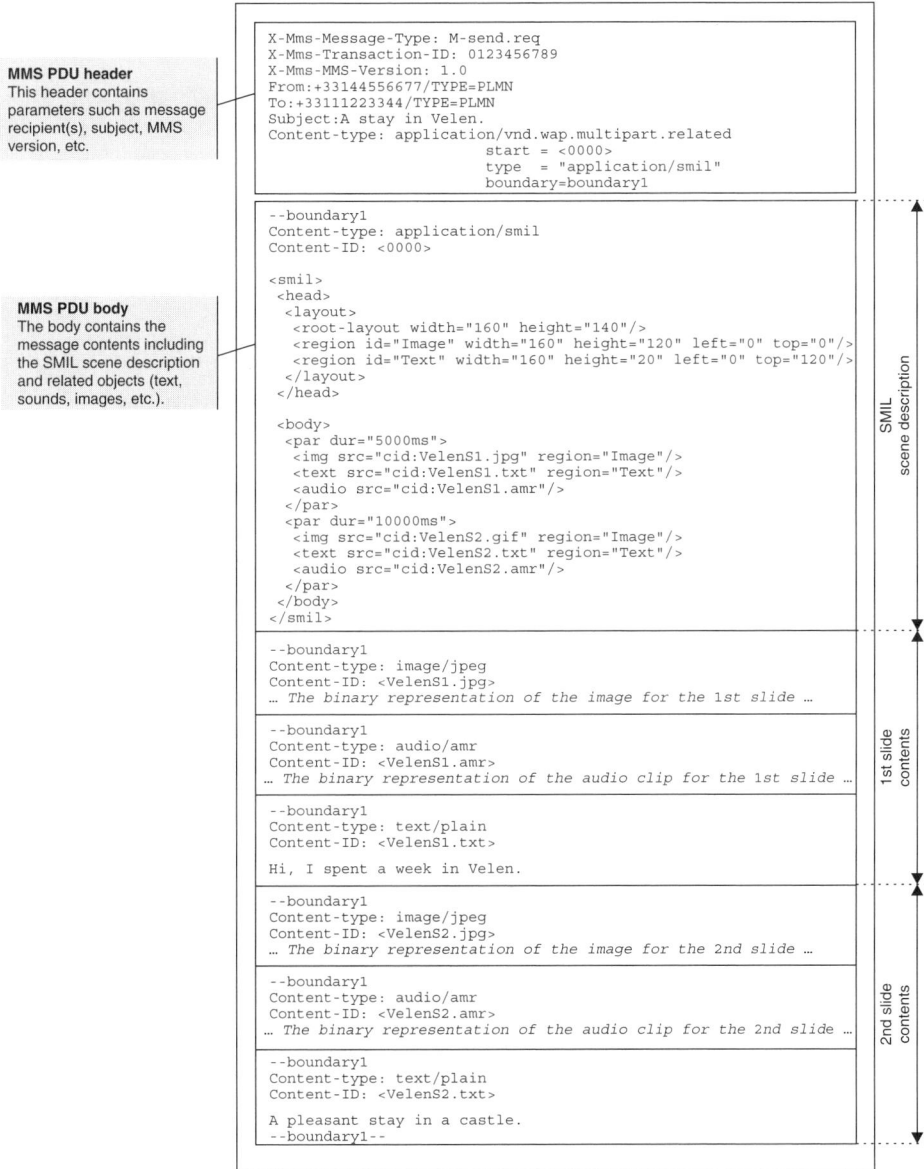

**Figure 5.11**   Full multimedia message example

message composed of two slides.[5] Each slide contains one image, some text and one sound clip. The size of the entire MMS PDU is 25 kB.

Figure 5.12 shows how the message may look like when displayed on the receiving device.

---

[5] The binary representation of the message is available from this book's companion website at http://www.lebodic.net/mms_resources.htm.

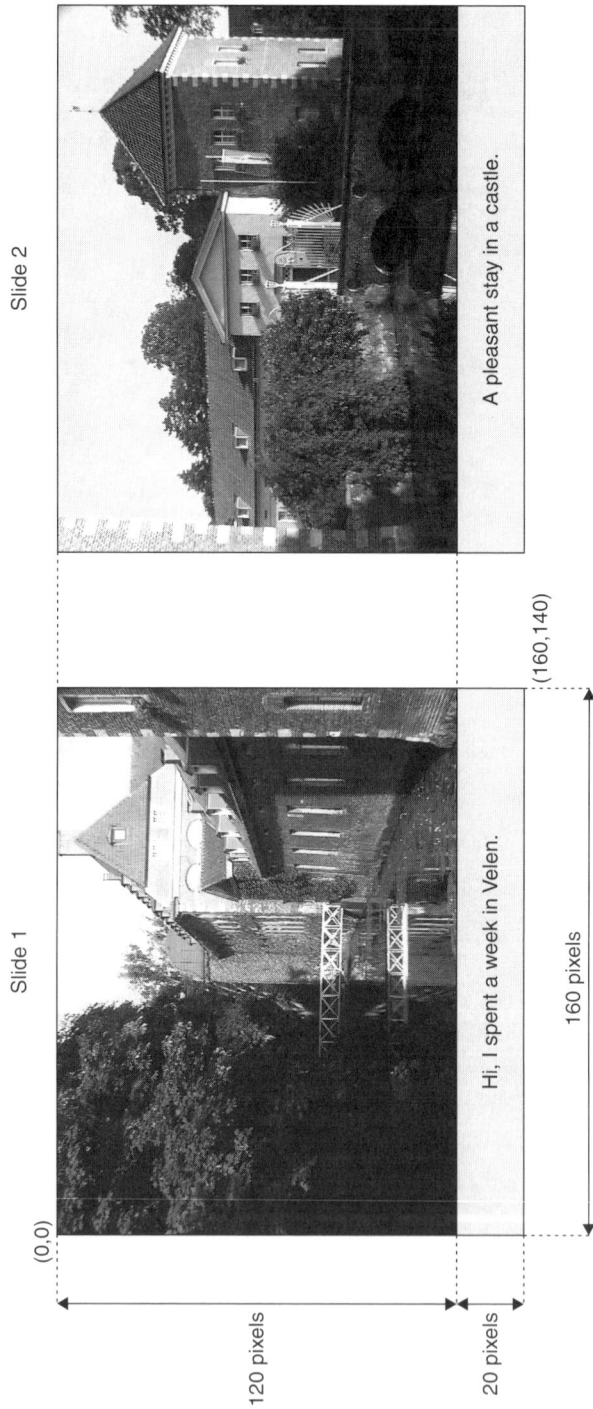

**Figure 5.12**  Message rendering on receiving device

## 5.6 Forward-lock of Media Objects

In the content-to-person scenario, certain business models require the capability to prevent users from redistributing freely or modifying protected contents. In the context of MMS, preventing such usages of protected media objects is performed by encapsulating the media object in an OMA DRM forward-lock message, also known as envelope (see introduction to OMA digital rights management in Section 3.6).

Each media object of a message that needs to be protected is encapsulated into a single forward-lock envelope. A forward-lock envelope is organized as a one-part structure for which the encapsulated element is the protected media object as shown below:

```
Content-type: application/vnd.oma.drm.message;
boundary = boundary_mms
```

As for multipart structures, the boundary delimiter of the forward-lock content type is used for delimiting the encapsulated element.

The whole multimedia message cannot be encapsulated in a forward-lock envelope. Figure 5.13 shows a multimedia message in which the image of the first slide is encapsulated in a forward-lock envelope.

Note that the OMA DRM forward-lock technique offers a simple mechanism for protecting media objects against forwarding and modifications in a fairly secure environment. In the Internet world, such a technique would be easily bypassed by any knowledgeable software developer. However, MMS device resources are not yet completely open to applications developed by parties other than the device vendors and their trusted partners. In this context, it is consequently more difficult to bypass the OMA DRM forward-protection.

It was shown in Section 3.6 that OMA DRM is based on three content protection techniques: combined delivery, separate delivery and forward-lock (a special case of combined delivery). At the time of writing, only OMA DRM forward-lock was applicable to MMS. However, it is expected that standards will evolve in order to provide support for combined delivery and separate delivery for MMS in the near future.

## 5.7 Message Size Measurement

A definition for message size is difficult to derive for multimedia messages. Indeed, a multimedia message can take various forms according to the interface it is conveyed over. Content adaptation can also change the size of a retrieved message compared to the size of the submitted message.

However, it is fundamental for the success of MMS to rely on a commonly understood message size definition since it is used for charging purpose, for characterizing a message in a notification, and so on. In the past, standardization organizations published various methods for calculating the size of a message. Only recently,

**MMS PDU header**
This header contains parameters such as message recipient(s), subject, MMS version, etc.

```
X-Mms-Message-Type: M-retrieve.conf
X-Mms-Transaction-ID: 0123456789
X-Mms-MMS-Version: 1.0
Message-ID: 0066
From: +33144556677/TYPE=PLMN
To: +33111223344/TYPE=PLMN
Date: Wed Jan 23 09:45:33 2002
Subject: A stay in Velen.
Content-type: application/vnd.wap.multipart.related;
              start = <0000>;
              type  = "application/smil";
              boundary=boundary_mms
```

**MMS PDU body**
The body contains the message contents including the SMIL scene description and related media objects (text, audio clips, images, etc.).

```
--boundary_mms
Content-type: application/smil
Content-ID: <0000>

<smil>
 <head>
  <layout>
   <root-layout width="160" height="140"/>
   <region id="Image" width="160" height="120" left="0" top="0"/>
   <region id="Text" width="160" height="20" left="0" top="120"/>
  </layout>
 </head>

 <body>
  <par dur="5000ms">
   <img src="cid:VelenS1.dm" region="Image"/>
   <text src="cid:VelenS1.txt" region="Text"/>
  </par>
 </body>
</smil>
```

**DRM message/envelope**
The DRM forward-lock envelope contains the protected media object.

```
--boundary_mms

Content-type: application/vnd.oma.drm;
                       boundary=boundary_drm
Content-ID: <VelenS1.dm>

--boundary_drm
```

Jpeg image encapsulated in an OMA DRM forward-lock envelope.

```
Content-type: image/jpeg
Content-transfer-encoding: binary

… The binary representation of the image for the 1st slide …

--boundary_drm--
```

```
--boundary_mms

Content-type: text/plain
Content-ID: <VelenS1.txt>

Hi, I spent a week in Velen.

--boundary_mms--
```

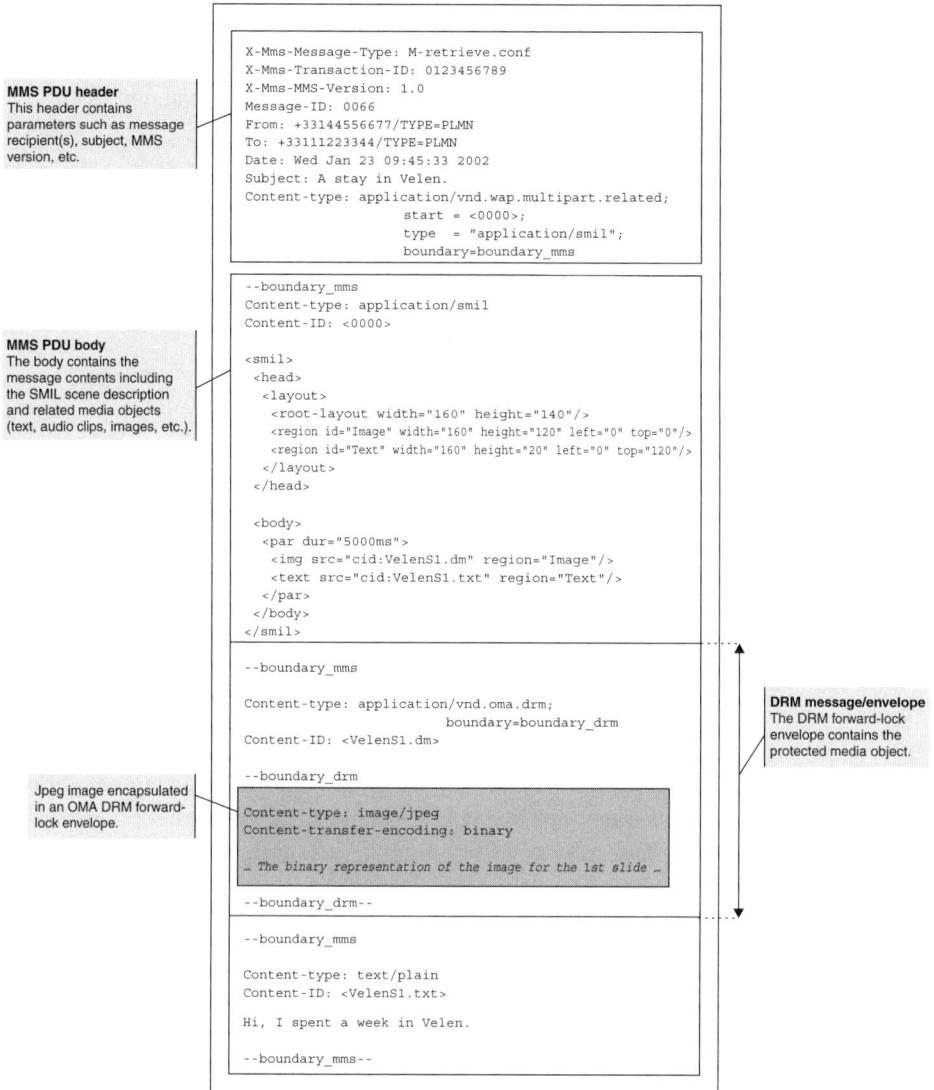

**Figure 5.13**   Multimedia message with forward protected media object

3GPP published a method that seems to have won wide acceptance ([3GPP 23.140] Release 5). The definition is as follows:

*Message size: the size of a multimedia message is calculated as the sum of the size of the message subject and the size of all elements (including scene description) encapsulated in the message body parts.*

---

**BOX 5.7 Relationships between message distribution indicator and OMA DRM forward-lock**

The application of OMA DRM forward-lock, in the context of MMS, is introduced in standards from MMS 1.2. The concept of message distribution indicator was also introduced in standards from MMS 1.2. Note that these two concepts are not competing technologies for DRM. The message distribution indicator is simply an 'informational' indication provided to the user stating whether or not the VAS Provider (VASP), originating the message, requested the message not to be forwarded by the recipient(s). The standard does not mandate any functional requirements on the recipient device for preventing the redistribution (or modification) of the associated media object(s)/message. On the other hand, a device compliant to OMA DRM standards is expected to prevent the user from forwarding (or modifying) any media object encapsulated in an OMA-DRM forward-lock envelope.

---

The size of the message subject is the size of the value assigned to the subject parameter without the subject parameter name (textual representation) or corresponding assigned number (binary representation). For instance, if the subject is defined as 'subject: A small house' (textual representation), then only 'A small house' is taken into consideration for the size calculation.

The size of an element encapsulated in a message body part is either of the following:

- The size of the media object (e.g. image, video clip, etc.) if the content type of the message body part is not a multipart structure (multipart/related or multipart/mixed). Body part parameters (e.g. content-type, content-location, etc.) are not taken into account for the calculation of the size of a message element.
- The size of an element that is a multipart/one-part structure (e.g. protected media object, see Section 5.6) is the total number of bytes contained in the multipart element contained in the main body part excluding parameters of the main body part but including parameters of the inner multipart element(s).

## Further Reading

I.E.G. Richardson, *Video Codec Design*, John Wiley & Sons, Chichester, 2002.

D.C.A. Bulterman, (Ed. Peiya Liu), SMIL 2.0, part I: overview, concepts and structure, *IEEE Multimedia Magazine*, **8**(4), 2001, 82–88.

D.C.A. Bulterman, (Ed. Peiya Liu), SMIL 2.0, part II: examples and comparisons, *IEEE Multimedia Magazine*, **9**(1), 2002, 74–84.

# 6

# Transactions Flows

Each end-to-end feature offered by the Multimedia Messaging Service (MMS) relies on a series of consecutive transactions occurring over one or more of the eight identified MMS interfaces. These transactions allow the transfer of messages and associated reports between MMS communicating entities including network servers and mobile devices.

This chapter presents transactions required for the realization of the service in the context of two representative scenarios: the person-to-person and content-to-person scenarios. Furthermore, this chapter provides a comprehensive description of transactions than can occur over MM1, MM4 and MM7 interfaces. The transport protocols for these interfaces have been fully defined by standardization organizations. Other relevant interfaces are also outlined.

## 6.1 Introduction to the MMS Transaction Model

MMS entities (MMS clients or MMS centres) communicate by invoking transactions over a set of eight interfaces. A transaction is typically composed of a service *request* and a corresponding service *response/confirmation* containing the transaction results (e.g. message sending request and message sending confirmation). However, several transactions are limited to a service request only and are also known as *indications*. A Protocol Data Unit (PDU) is associated with each service request or response that can occur over one of the MMS interfaces. A PDU is composed of a set of mandatory, optional or conditional parameters.

In this book, the Third Generation Partnership Project (3GPP) convention for naming PDUs is employed for the description of transactions that can occur over MM4 and MM7 interfaces. The convention consists of suffixing the request name by .REQ and suffixing the response name by .RES. In addition, request and response names are prefixed by the associated interface name. For instance, the request for routing forward a message over the MM4 interface is named MM4_forward.REQ and the corresponding response is named MM4_forward.RES. On the other hand, the Open Mobile Alliance (OMA) convention for naming PDUs is employed for the description of transactions that can occur over the MM1 interface. This convention is further presented in Section 6.2.

*Multimedia Messaging Service: An Engineering Approach to MMS* Gwenaël Le Bodic
© 2003 John Wiley & Sons, Ltd ISBN: 0-470-86253-X

### 6.1.1 Person-to-person Scenarios

The simplest transaction flow for a message exchange is the one that involves two MMS clients attached to the same MMS Environment (MMSE). In this context, the message exchange usually involves one MMSC only. The process of exchanging a message is composed of four steps as shown in Figure 6.1 and described below:

1. *Message submission over the originator MM1 interface*: This transaction is composed of a submission request (`M-send.req` PDU) and a corresponding submission response (`M-send.conf` PDU).
2. *Message notification over the recipient MM1 interface*: This transaction is composed of a notification indication (`M-notification.ind` PDU) and a notification response indication (`M-notifyresp.ind` PDU).
3. *Message retrieval over the recipient MM1 interface*: This transaction is composed of a retrieval request (`WSP/HTTP GET.req` PDU), retrieval response (`M-retrieve.conf` PDU) and optionally a retrieval acknowledgement indication (`M-acknowledge.ind` PDU).
4. *Delivery reporting over the originator MM1 interface*: A delivery report is conveyed over the MM1 interface only if the generation of a delivery report was requested during message submission and if the recipient did not deny the generation of such a report. The transaction is composed of a delivery-report indication (`M-delivery.ind` PDU).

The exchange of a message between MMS clients attached to distinct MMSEs involves two MMSCs: the originator MMSC and the recipient MMSC. The two MMSCs are interconnected via the MM4 interface. The process of exchanging a message is composed of four to six steps as shown in Figure 6.2 and described below:

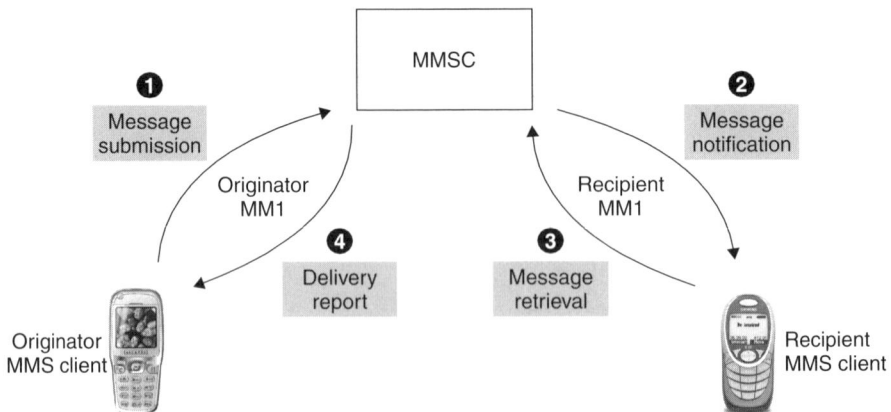

**Figure 6.1**   General transaction flow/Message exchange – single MMSE

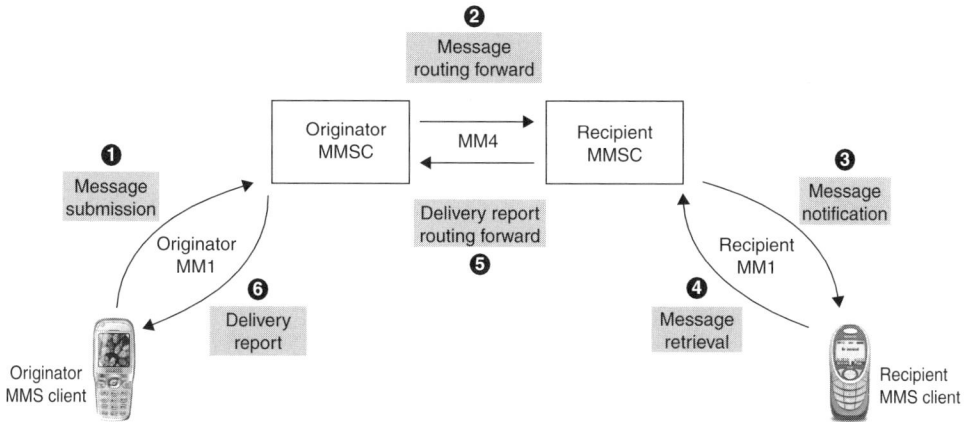

**Figure 6.2**  General transaction flow/Message exchange – multiple MMSEs

1. *Message submission* over the originator MM1 interface.
2. *Routing forward the message* over the MM4 interface: This transaction is composed of a forward request (MM4_forward.REQ PDU) and a corresponding forward response (MM4_forward.RES PDU).
3. *Message notification* over the recipient MM1 interface.
4. *Message retrieval* over the recipient MM1 interface.
5. *Routing forward the delivery report* over the MM4 interface: This transaction is composed of a submission request (MM4_delivery_report.REQ PDU) and a corresponding submission response (MM4_delivery_report.RES PDU). The delivery report is generated by the recipient MMSC upon confirmation of message retrieval by the recipient MMS client.
6. *Delivery reporting* over the originator MM1 interface: A delivery report is conveyed over the MM1 interface only if the generation of a delivery report was requested during message submission.

Note that a read report can also be generated when the message has been read by the recipient. For the sake of clarity, transactions related to the management of a read report are not shown in Figure 6.1 and Figure 6.2. A read report is generated by the recipient MMS client when the recipient has read or deleted the message only if the generation of a read report was requested by the originator and if the recipient did not deny the generation of such a report. In the recipient environment, the read report is conveyed from the recipient MMS client to the recipient MMSC in the form of a read-report indication (M-read-rec.ind PDU) over the MM1 interface. The read report is forwarded over the MM4 interface with a transaction composed of a forward request (MM4_read_reply_report.REQ PDU) and a forward response (MM4_read_reply_report.RES PDU). Eventually, the read report is pushed to the originator MMS client as part of a read-report indication (M-read-orig.ind PDU) in the originator MMSE over the MM1 interface.

### 6.1.2 Content-to-person Scenarios

*Content to person* refers to the scenario where the message originates from a VAS application and is delivered to one or more MMS recipients. For this purpose, the VAS Provider (VASP) operates a VAS application, usually Internet hosted, and connected to an MMSC via the MM7 interface. In this configuration, the MMSC interacts with recipient MMS clients over the MM1 interface as already described in the previous section.

Figure 6.3 shows interactions between a VAS application, the MMSC and one recipient MMS client for the exchange of a message from the VAS application down to the recipient MMS client. Optionally, the VAS application can request the generation of delivery and read reports for the message. Note that the recipient often has the possibility of denying the generation of such reports.

The process of exchanging a message from the VAS application to the recipient MMS client is composed of four to six steps as described below:

1. *Message submission* over the MM7 interface. This transaction is composed of a submission request (`MM7_submit.REQ` PDU) and a corresponding submission response (`MM7_submit.RES` PDU).
2. *Message notification* over the MM1 interface (see previous section).
3. *Message retrieval* over the MM1 interface (see previous section).
4. *Delivery reporting* over the originator MM7 interface: A delivery report is conveyed over the MM7 interface only if the generation of a delivery report was requested during message submission and if the recipient did not deny the generation of such a report. The transaction is composed of a delivery-report request (`MM7_delivery_report.REQ` PDU) and a corresponding delivery-report response (`MM7_delivery_report.RES` PDU).
5. *Read reporting* over the *MM1* interface (see previous section).
6. *Read reporting* over the *MM7* interface: As for delivery reports, a read report is conveyed over the MM7 interface only if the generation of a read report was requested during message submission and if the recipient did not deny the generation of such a report. The transaction is composed of a read-report request (`MM7_read_reply_report.REQ` PDU) and a corresponding read-report response (`MM7_read_reply_report.RES` PDU).

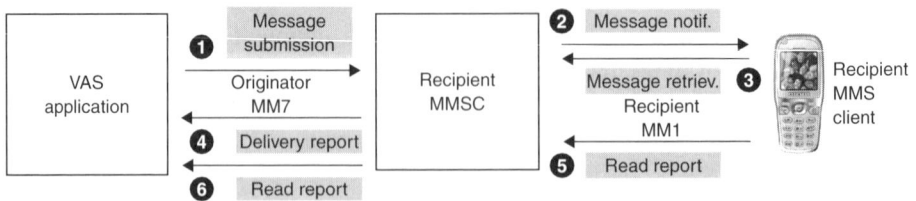

**Figure 6.3** General transaction flow/Message exchange – from VAS application to MMS client(s)

There is always a period during which the submitted message is stored temporarily in the MMSC, awaiting retrieval from the recipient MMS client. During this period, it is possible for the originating VAS application to cancel the message delivery or to replace the message due to be delivered. The cancel transaction over the MM7 interface is composed of a cancel request (`MM7_cancel.REQ` PDU) and a corresponding cancel response (`MM7_cancel.RES` PDU). Similarly, the replace transaction over the MM7 interface is composed of a replace request (`MM7_replace.REQ` PDU) and a corresponding replace response (`MM7_replace.RES` PDU).

On the other way round, the MMS client also has the possibility of submitting a message addressed to a VAS application. This may be a reply related to a previous message originated by the VAS application or it can be a new unrelated message. Figure 6.4 shows interactions between the MMS client, the MMSC and the VAS application for the exchange of a message from the MMS client up to the VAS application.

The process of exchanging a message from the MMS client to the VAS application is composed of two to three steps as described below:

1. *Message submission* over the originator MM1 interface (see Section 6.1.1).
2. *Message delivery* over the MM7 interface. This transaction is composed of a delivery request (`MM7_deliver.REQ` PDU) and a corresponding delivery response (`MM7_deliver.RES` PDU).
3. *Delivery reporting* over the MM1 interface (see Section 6.1.1).

### 6.1.3 How to Read the PDU Description Tables

This chapter details the composition of each PDU that can occur over several MMS interfaces. The composition of a PDU is represented in a concise table as shown in Figure 6.5.

Values of the 'From' column indicate from which version of the MMS standards the corresponding parameter becomes applicable. For instance, the value '1.1' means that the corresponding parameter was introduced in MMS 1.1 and therefore such a parameter may be found in MMS PDUs versioned 1.1, 1.2 and so on but will never be present in a PDU versioned 1.0.

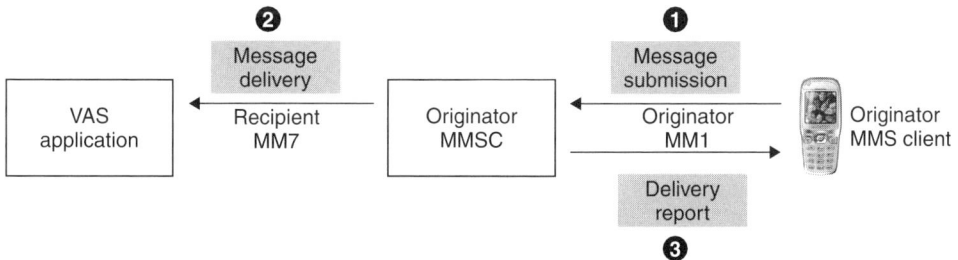

**Figure 6.4** General transaction flow/Message exchange – from MMS client to VAS application

**Figure 6.5**   How to read the PDU description tables

The column 'St.' which stands for 'Status', indicates whether the presence of the corresponding parameter is mandatory, optional or conditional as shown below:

● mandatory.
○ optional.
© conditional.

## 6.2 MM1 Interface, MMS Client – MMSC

The MM1 interface allows interactions between the MMS client and the MMSC. Several primitives, also known as PDUs, can be invoked over this interface for transactions such as message notification, message submission, message retrieval, and so on. According to the OMA terminology for MMS [OMA-MMS-Arch], the MM1 interface is also known as the $MMS_M$ interface.

In this book, the convention for naming MM1 PDUs is the one used in OMA standards. This convention consists of prefixing the primitive name with the interface name and suffixing the primitive name with the primitive type as shown in Figure 6.6.

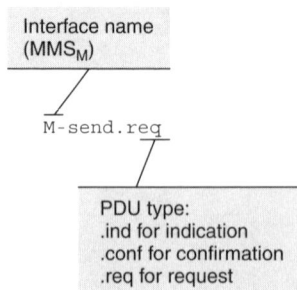

**Figure 6.6**   Convention for naming MM1 PDUs

Three types of PDUs can be invoked over the MM1 interface:

- *Request*: A request PDU is invoked from an MMS entity (MMS client or MMSC) to request a service to be provided by another MMS entity (MMS client or MMSC). In this context, the serving MMS entity, which accepts or rejects the request, notifies the requesting MMS entity of the request status with a confirmation PDU. A request PDU is often marked with a transaction identifier. The name of a request PDU is suffixed with .req (e.g. M-send.req).
- *Confirmation/response*: A confirmation PDU is invoked by a serving MMS entity to confirm the status of a previously requested service. It is marked with the transaction identifier of the corresponding request PDU. The name of a confirmation PDU is suffixed with .conf (e.g. M-send.conf).
- *Indication*: An indication PDU is invoked by an MMS entity to notify another MMS entity of the occurrence of an event (message notification, reports, etc.). The name of an indication PDU is suffixed with .ind (e.g. M-notification.ind). An indication is not confirmed.

Table 6.1 lists the ten transactions that can occur over the MM1 interface (20 PDUs).

For a transaction composed of a request and a corresponding confirmation, the confirmation usually contains a status code indicating if the serving MMS entity accepts or rejects the service request. In the case of rejection, the confirmation status code identifies the reason for rejection. It is common to differentiate rejections due to permanent errors from rejections due to transient errors. A rejection due to a *permanent error* means that the request in its current form is not acceptable, whereas a rejection due to a *transient error* means that the request cannot be processed because of some transient error conditions, but the request may be accepted, in its current form, later. For instance, an MMS client failing to submit a message to an MMSC because of a transient error (e.g. MMSC busy) may reattempt to request the service in the same way at a later stage.

The next sections of this chapter present interactions between MMS entities for each transaction flow (e.g. message submission, message retrieval, etc.). MMS PDUs involved in these transaction flows are composed of a PDU header (a sequence of parameters) and an optional PDU body (containing multimedia message contents or informational text) as shown in Figure 6.7. Each PDU is conveyed over the MM1 interface in a binary form for transport efficiency. Section 6.2.11 of this chapter explains how each parameter is binary encoded according to the type of its assigned values (integer, date, string, etc.) and provides guidelines regarding the maximum lengths of several parameter values. Once binary encoded, each MMS PDU is encapsulated in a transport-level Wireless Session Protocol (WSP) or HTTP PDU (request/response) and conveyed over a data connection established between the mobile device and the network or conveyed as part of an SMS message (WAP push). Figure 6.7 shows how an MMS PDU is encapsulated in a WSP or a HTTP PDU.

**Table 6.1**　List of MM1 transactions/PDUs

| Transaction | PDU name | Description | From OMA |
|---|---|---|---|
| Message submission | `M-send.req` | Message submission request | 1.0 |
| | `M-send.conf` | Message submission confirmation | |
| Notification | `M-notification.ind` | Message notification indication | 1.0 |
| | `M-notifyresp.ind` | Message notification response indication | |
| Message retrieval | `WSP/HTTP GET.req` | Message retrieval request | 1.0 |
| | `M-retrieve.conf` | Message retrieval confirmation | |
| | `M-acknowledge.ind` | Message retrieval acknowledgement indication | |
| Delivery report | `M-delivery.ind` | Delivery report indication | 1.0 |
| Read report | `M-read-rec.ind` | Read-report indication (recipient MMSE) | 1.1 |
| | `M-read-orig.ind` | Read-report indication (originator MMSE) | |
| Message forward | `M-forward.req` | Message forward request | 1.1 |
| | `M-forward.conf` | Message forward confirmation | |
| Message store or update into MMBox | `M-Mbox-Store.req` | MMBox message store/update request | 1.2 |
| | `M-Mbox-Store.conf` | MMBox message store/update confirmation | |
| View contents of MMBox | `M-Mbox-View.req` | MMBox contents view request | 1.2 |
| | `M-Mbox-View.conf` | MMBox contents view confirmation | |
| Message upload to MMBox | `M-Mbox-Upload.req` | MMBox message upload request | 1.2 |
| | `M-Mbox-Upload.conf` | MMBox message upload confirmation | |
| Message deletion from MMBox | `M-Mbox-Delete.req` | MMBox message deletion request | 1.2 |
| | `M-Mbox-Delete.conf` | MMBox message deletion confirmation | |

At the WSP/HTTP layer, an MMS PDU is marked with the following content type:

```
application/vnd.wap.mms-message
```

Each MMS PDU header is composed of a number of optional, conditional and mandatory parameters. If an optional parameter is omitted from the PDU header, then a default value is implicitly considered for this parameter. In this book, the default value is indicated in the corresponding parameter description when appropriate.

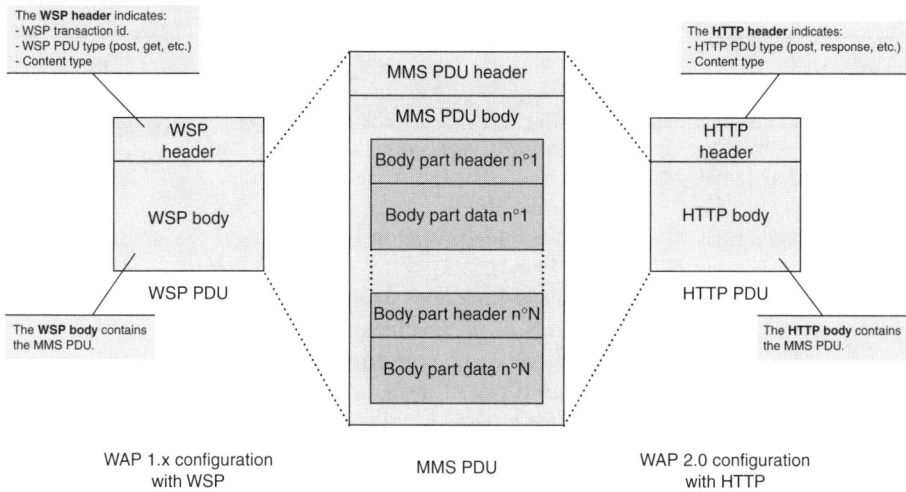

The **WSP header** indicates:
- WSP transaction id.
- WSP PDU type (post, get, etc.)
- Content type

The **HTTP header** indicates:
- HTTP PDU type (post, response, etc.)
- Content type

WSP header

WSP body

WSP PDU

The **WSP body** contains the MMS PDU.

MMS PDU header

MMS PDU body

Body part header n°1

Body part data n°1

Body part header n°N

Body part data n°N

HTTP header

HTTP body

HTTP PDU

The **HTTP body** contains the MMS PDU.

WAP 1.x configuration with WSP

MMS PDU

WAP 2.0 configuration with HTTP

**Figure 6.7**  Encapsulation of a MMS PDU in a WSP or HTTP PDU

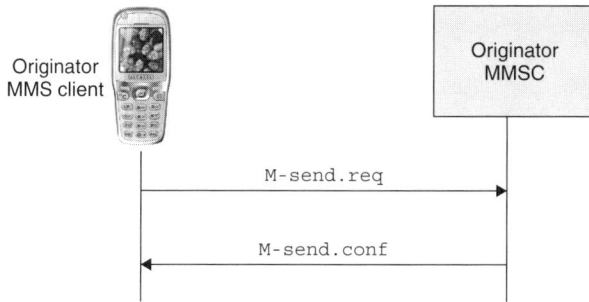

Originator MMS client

Originator MMSC

M-send.req

M-send.conf

**Figure 6.8**  MM1 message submission transaction flow

## 6.2.1 Message Submission

Message submission refers to the submission of a multimedia message from the originator MMS client to the originator MMSC. The transaction flow for message submission is shown in Figure 6.8. The PDU corresponding to the submission request is named M-send.req and is confirmed by the PDU named M-send.conf.

In WAP technical realizations, the mobile device establishes a data connection with the network. This allows the MMS client to communicate with the MMSC. This data connection can rely on a circuit-switched data connection (e.g. GSM) or a packet-switched connection (e.g. GPRS). The connection is typically established through a WAP gateway in a WAP 1.x configuration. It is common for operators to set up a dedicated WAP gateway for handling MMS traffic in addition to the one that handles browsing and download traffics.

Parameters composing the message submission request PDU (M-send.req) are presented in Table 6.2, whereas parameters composing the message submission

**Table 6.2** MM1 message submission request (`M-send.req`)

| Parameter name | Description | From OMA | St. |
|---|---|---|---|
| `X-Mms-Message-Type` | MMS PDU type. Value: `M-send-req` | 1.0 | ● |
| `X-Mms-Transaction-ID` | Unique identifier for the submission transaction. | 1.0 | ● |
| `X-Mms-MMS-Version` | MMS protocol version such as 1.0, 1.1 or 1.2. | 1.0 | ● |
| `Date` | Date and time of message submission. | 1.0 | ○ |
| `From` | Address of the originator MMS client (phone number or email address) or 'insert token' if the originator address is to be provided by the MMSC. | 1.0 | ● |
| `To` | One or multiple addresses (phone number or email address) for message recipient(s). Primary recipients. | 1.0 | ○ |
| `Cc` | One or multiple addresses (phone number or email address) for message recipient(s). Secondary recipients. | 1.0 | ○ |
| `Bcc` | One or multiple addresses (phone number or email address) for message recipient(s). Secondary recipients/ blind copy. | 1.0 | ○ |
| `Subject` | A short textual description for the message. | 1.0 | ○ |
| `X-Mms-Message-Class` | Message class such as 'auto' (automatically generated by the MMS client), 'personal' (default), 'advertisement' and 'informational'. Other classes can also be defined in the form of text strings. | 1.0 | ○ |
| `X-Mms-Expiry` | Expiry date. Default value for this parameter is 'maximum'. | 1.0 | ○ |
| `X-Mms-Delivery-Time` | Earliest delivery time. Default value for this parameter is 'immediate delivery'. | 1.0 | ○ |
| `X-Mms-Priority` | Priority such as 'low', 'normal' (default) or 'high'. | 1.0 | ○ |
| `X-Mms-Sender-Visibility` | Visibility of the sender address. This parameter is either set to 'show' (default) for showing the sender address to recipient(s) or 'hide' for hiding the sender address to recipient(s). From MMS 1.2, 'show' is not anymore the default value for this parameter. If this parameter is not present in an MMS 1.2 PDU, then network preferences for the sender anonymity feature are used. | 1.0 | ○ |

**Table 6.2**   (*continued*)

| Parameter name | Description | From OMA | St. |
|---|---|---|---|
| X-Mms-Delivery-Report | Request for a delivery report. This parameter indicates whether or not delivery report(s) are to be generated for the submitted message. Two values can be assigned to this parameter: 'yes' (delivery report is to be generated) or 'no' (no delivery report requested). If the message class is 'auto', then this parameter is present in the submission PDU and is set to 'no'. | 1.0 | ○ |
| X-Mms-Read-Report | Request for a read report. This parameter indicates whether or not read reports are to be generated for the message. Two values can be assigned to this parameter: 'yes' (read report is to be generated) or 'no' (no read report requested). If the message class is auto, then this parameter is present in the submission PDU and is set to 'no'. | 1.0 | ○ |
| X-Mms-Reply-Charging | Request for reply charging. The presence of this parameter indicates that reply charging is requested by the message originator. Two values can be assigned to this parameter: 'requested' when the originator is willing to pay for the message reply(s) or 'requested text only' when the originator is willing to pay for message reply(s) containing text only. In any case, two parameters (reply message size and reply deadline) specify conditions for the message reply to be paid for by the originator. | 1.1 | ○ |
| X-Mms-Reply-Charging-Deadline | Reply charging – deadline. This parameter specifies the latest time for the recipient(s) to submit a message reply. This parameter is only present in the PDU if reply charging is requested. | 1.1 | ○ |
| X-Mms-Reply-Charging-Size | Reply charging – maximum message size. This parameter specifies the maximum size for message replies. This parameter is only present in the PDU if reply charging is requested. | 1.1 | ○ |

(*continued overleaf*)

**Table 6.2** (*continued*)

| Parameter name | Description | From OMA | St. |
|---|---|---|---|
| X-Mms-Reply-Charging-ID | Reply charging – identification. This parameter is inserted in a reply message only and refers to the original message identifier (Message-ID parameter). | 1.1 | ○ |
| X-Mms-Store | MMBox storage request. This parameter indicates whether the originator MMS client requests to save the message in the originator's MMBox in addition to sending it. | 1.2 | ○ |
| X-Mms-MM-State | MMBox message state. When X-Mms-Store is set to 'yes', this parameter indicates the message state in the originator's MMBox (e.g. sent, draft, etc.). If X-Mms-Store is set to 'yes' and if this parameter is not present then the message default state is 'sent'. | 1.2 | ○ |
| X-Mms-MM-Flags | MMBox message flag. This parameter indicates the list of flags associated to a message stored in the MMBox (considered only if X-Mms-Store is set to 'yes'). | 1.2 | ○ |
| Content-Type | Content type of the multimedia message. (e.g. application/vnd.wap. multipart.related). | 1.0 | ● |

confirmation PDU (M-send.conf) are presented in Table 6.3. The body of the submission request PDU contains the submitted multimedia message and the PDU header represents the message envelope. The submission confirmation PDU does not contain any body (header only).

The originator MMS client can specify a date and time for the submission request (parameter Date of the request in the GMT[1] format). If such date and time are not provided by the MMS client, then it becomes the originator MMSC's responsibility to provide the date and time for the submitted message. In any case, the MMSC always has the possibility of overwriting a date and time provided by the MMS client. Note that, unfortunately, the MMS protocol does not provide any means for the MMS client to be informed of the date assigned by the originator MMSC. Each one of the addressing parameters may appear multiple times.

Recipient addressing parameters (To, Cc and Bcc) are all optional. However, at least one recipient address shall be provided by the MMS client to the MMSC.

---

[1] Greenwich Mean Time.

**Table 6.3**  MM1 message submission confirmation (`M-send.conf`)

| Parameter name | Description | From OMA | St. |
|---|---|---|---|
| `X-Mms-Message-Type` | MMS protocol data unit type. Value: `M-send-conf` | 1.0 | ● |
| `X-Mms-Transaction-ID` | Unique identifier for the submission transaction. The same as the one for the corresponding submission request. | 1.0 | ● |
| `X-Mms-MMS-Version` | MMS protocol version such as 1.0, 1.1 or 1.2. | 1.0 | ● |
| `X-Mms-Response-Status` | Status code for the submission transaction. The submission request can be accepted or rejected (permanent or transient errors). See status codes in Appendix B. | 1.0 | ● |
| `X-Mms-Response-Text` | Human readable description of the transaction status. | 1.0 | ○ |
| `Message-ID` | Message unique identifier. This identifier is always provided by the MMSC if the submission request is accepted. | 1.0 | ○ |
| `X-Mms-Content-Location` | Reference to the message stored in the MMBox. This parameter is present only if the three following conditions are fulfilled: <br> – the originator MMSC supports the MMBox feature, <br> – the `X-Mms-Store` parameter was present in the corresponding submission request, <br> – the `X-Mms-Store-Status` indicates 'success'. <br> When available, this parameter provides a reference to the message stored in the MMBox (reference used later for message retrieval or view request). | 1.2 | ○ |
| `X-Mms-Store-Status` | MMBox message status. This parameter is present only if the two following conditions are fulfilled: <br> – the originator MMSC supports the MMBox feature, <br> – the `X-Mms-Store` parameter was present in the corresponding submission request. <br> When available, this parameter indicates whether or not the submitted message has been successfully stored in the MMBox. See status codes in Appendix D. | 1.2 | ○ |
| `X-Mms-Store-Status-Text` | MMBox message textual status. Textual description qualifying the value assigned to the `X-Mms-Store-Status` parameter. | 1.2 | ○ |

The client has the possibility of requesting reply charging for the submitted message (from MMS 1.1). However, reply charging may not be supported by the MMSC. If reply charging is not supported, then the MMSC rejects the submission request by providing a confirmation with the appropriate error code.

If the submission request is accepted by the MMSC, then the MMSC provides a unique message identifier (parameter `Message-ID` of the confirmation). This identifier can later be used for correlating reports and message replies (reply charging) with the original message.

In the event of the submission request not being accepted by the MMSC, the MMSC specifies the reason for rejection in the confirmation. The reason may be of permanent nature (e.g. message badly formatted) or of transient nature (e.g. MMSC is temporarily unavailable).

If the Multimedia Message Box (MMBox) concept is not supported by the originator MMSC, then the submission request MMBox-related parameters (`X-Mms-Store`, `X-Mms-MM-State` and `X-Mms-MM-Flags`) are ignored by the MMSC.

The submission request PDU is transferred from the MMS client to the MMSC. It is conveyed as part of a WSP/HTTP Post request. Figure 6.9 shows the partial hexadecimal representation of a submission request PDU with the following characteristics:

```
X-Mms-Message-Type: M-send.req
X-Mms-Transaction-ID: 0123456789
X-Mms-MMS-Version: 1.0
From: +33144556677/TYPE = PLMN
To: +33111223344/TYPE = PLMN
Subject: A stay in Velen.
Content-type: application/vnd.wap.multipart.related
            start = <0000>
            type = ''application/smil''
```

Contents of the message (partially represented in Figure 6.9) comprise a SMIL scene description (`Content-ID: <0000>`). This scene description represents two slides,

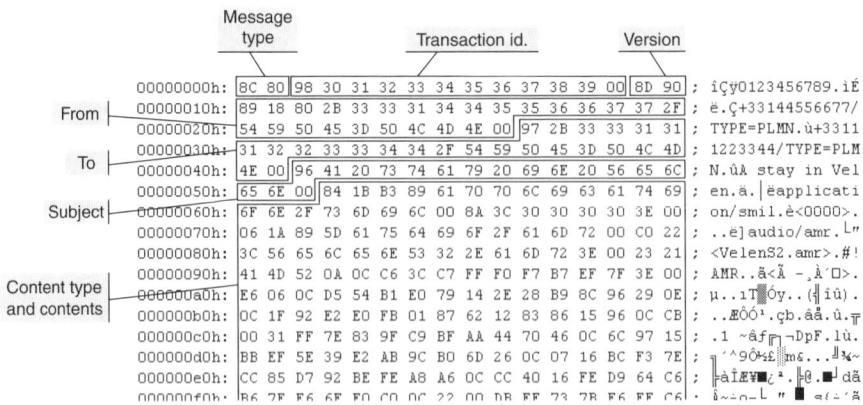

**Figure 6.9** Hexadecimal trace dump for submission request (`M-send.req`)

each of them containing an image, an Adaptive Multi-Rate (AMR) sound and a short text. The size of the entire MMS PDU is 25 kB and a binary file containing this PDU is available from this book's companion website.

### 6.2.2 Message Notification

Once a message submission has been accepted by the originator MMSC, the originator MMSC analyses message recipient addresses and identifies corresponding recipient MMSCs. The message is routed forward to all recipient MMSCs. Note that if originator and recipient MMS clients are attached to the same MMSE, then the MMSC plays the role of both originator and recipient MMSCs. Upon receipt of the message, the recipient MMSC builds a compact notification describing the message (class, size, priority, etc.) and delivers it to the recipient MMS client. With the notification, the recipient MMS client can retrieve the corresponding message immediately upon receipt of the notification (immediate message retrieval) or later at the user's own convenience (deferred message retrieval). Figure 6.10 shows the transaction flow for the delivery of a notification from the recipient MMSC to the recipient MMS client.

Parameters composing the notification indication PDU (M-notification.ind) are presented in Table 6.4. The notification indication PDU does not contain any body (header only).

Upon receipt of the notification, the MMS client can perform the following actions:

- *Message rejection*: The message is rejected without being retrieved by the recipient MMS client.
- *Immediate message retrieval*: The message is immediately retrieved without user explicit request.
- *Indication message awaiting retrieval*: The MMS client notifies the user that a message awaits retrieval. In this mode, the user can retrieve the message later at his/her own convenience (deferred message retrieval). With the notification, the user also has the possibility of forwarding the message to other recipients without

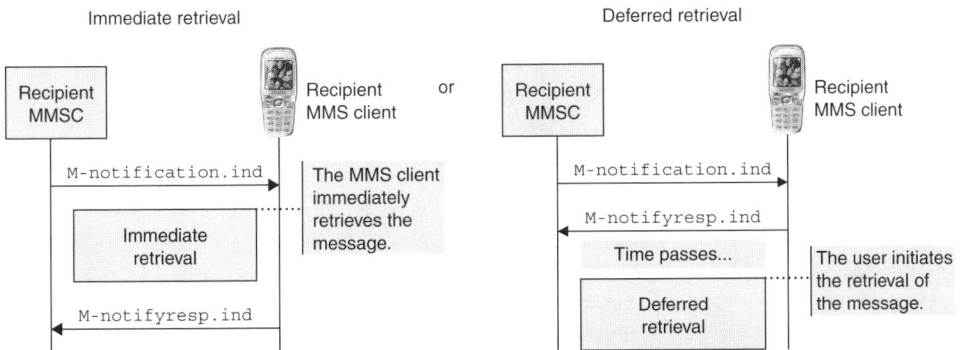

**Figure 6.10**   MM1 notification transaction flow

**Table 6.4**  MM1 notification indication (`M-notification.ind`)

| Parameter name | Description | From OMA | St. |
|---|---|---|---|
| X-Mms-Message-Type | MMS PDU type. Value: `M-notification-ind` | 1.0 | ● |
| X-Mms-Transaction-ID | Unique identifier for the notification transaction. | 1.0 | ● |
| X-Mms-MMS-Version | MMS protocol version such as 1.0, 1.1 or 1.2. | 1.0 | ● |
| From | Address of the originator MMS client (phone number or email address). This parameter is not present in the notification if the originator requested address hiding. | 1.0 | ○ |
| Subject | A short textual description for the message. | 1.0 | ○ |
| X-Mms-Message-Class | Message class such as 'auto' (automatically generated by the MMS client), 'personal' (default), 'advertisement' and 'informational'. Other classes can also be defined in the form of text strings. | 1.0 | ● |
| X-Mms-Expiry | Expiry date (in relative format). | 1.0 | ● |
| X-Mms-Priority | Priority such as 'low', 'normal' (default) or 'high'. | 1.2 | ○ |
| X-Mms-Delivery-Report | Request for a delivery report. This parameter indicates whether or not delivery reports are to be generated for the message. Two values can be assigned to this parameter: 'yes' (delivery report is to be generated) or 'no' (no delivery report requested). In MMS 1.0, the indication that the originator requested a delivery report is only present in the retrieved message. | 1.1 | ○ |
| X-Mms-Reply-Charging | Request for reply charging. The presence of this parameter indicates that reply charging is requested by the message originator. Two values can be assigned to this parameter: 'requested' when the originator is willing to pay for the message reply(s) or 'requested text only' when the originator is willing to pay for message reply(s) containing text only. In any case, two parameters (reply message size and reply deadline) specify conditions for the message reply to be paid for by the originator. | 1.1 | ○ |
| X-Mms-Reply-Charging-Deadline | Reply charging – deadline. This parameter specifies the latest time for the recipient to submit a message reply. This parameter is only present in the PDU if reply charging is requested. | 1.1 | ○ |
| X-Mms-Reply-Charging-Size | Reply charging – maximum message size. This parameter specifies the maximum size for message replies. This parameter is only present in the PDU if reply charging is requested. | 1.1 | ○ |

**Table 6.4**   (*continued*)

| Parameter name | Description | From OMA | St. |
|---|---|---|---|
| X-Mms-Reply-Charging-ID | Reply charging – identification. This parameter is only inserted in a notification corresponding to a reply message and refers to the original message identifier (Message-ID parameter). | 1.1 | ○ |
| X-Mms-Message-Size | Approximate size of the message. | 1.0 | ● |
| X-Mms-Stored | MMBox message availability. The value 'yes' assigned to this parameter indicates that the message is stored in the MMBox and is referenced by the value assigned to the X-Mms-Content-Location parameter. | 1.2 | ○ |
| X-Mms-Distribution-Indicator | Message distribution indicator. This parameter can be present in the notification when the message is sent by a VASP. The value 'no' indicates to the recipient that the originator requested the content of the message not to be distributed further. | 1.2 | ○ |
| X-Mms-Element-Descriptor | This parameter provides a description of the message main structure (e.g. multipart related, mixed, etc.). | 1.2 | ○ |
| X-Mms-Content-Location | Location of the message. This reference can be used by the MMS client for subsequent requests related to the corresponding message. | 1.0 | ● |

having to retrieve the message first (from MMS 1.1 only). The user can also instruct the MMS client to delete the notification. In this situation, the message remains in the MMSC message store until its validity period expires and will never be retrieved by the MMS client.

An important parameter of the notification indication PDU is the X-Mms-Content-Location parameter. The value assigned to this parameter indicates the location from which the message can be retrieved. An example of the message location that can be assigned to this parameter is

http://mmsc.operator.net/message-id-634

The protocol scheme, http or https, indicates whether or not a secure connection should be established for retrieving the corresponding message as explained in Section 4.14.

The X-Mms-Message-Size parameter of the notification indication provides an estimate of the message size. This estimated size may not represent exactly the size of the retrieved message. This is explained by the fact that message contents can be scaled down by the MMSC to meet recipient MMS client rendering capabilities just before the message is retrieved by the recipient MMS client. Furthermore, if the message originates from a value-added service provider, then the message can be replaced by a smaller or larger message between the delivery of the notification and the retrieval of the corresponding message. Consequently, the message size advertised in the notification is usually not used as a reason for automatic message rejection.

A notification indication PDU is conveyed from the recipient MMSC to the recipient MMS client as part of a WAP push ([OMA-WSP], see also Section 3.5.3). For this purpose, the PDU is either encapsulated into one or more SMS messages or, alternatively, conveyed as part of an already established data connection. At the transfer layer of the SMS protocol [3GPP-23.040], an SMS message is structured as a sequence of transfer protocol parameters (parameter names are prefixed with TP). For conveying an MMS notification, these parameters are assigned the following values:

TP-Protocol-Identifier: 0x00
TP-Data-Coding-Scheme: 8-bit data.
TP-User-Data: as shown in Figure 6.11.

An MMS push (notification or report) always has a push-application identification of 0x04 in the WSP header [OMA-MMS-Enc]. The notification indication PDU is binary encoded according to the rules described in Section 6.2.11. Figure 6.12 shows a hexadecimal trace dump for a notification indication PDU with the following characteristics:

X-Mms-Message-Type: M-notification.ind
X-Mms-Transaction-ID: Transaction-123456789-abcdefg
X-Mms-MMS-Version: 1.0

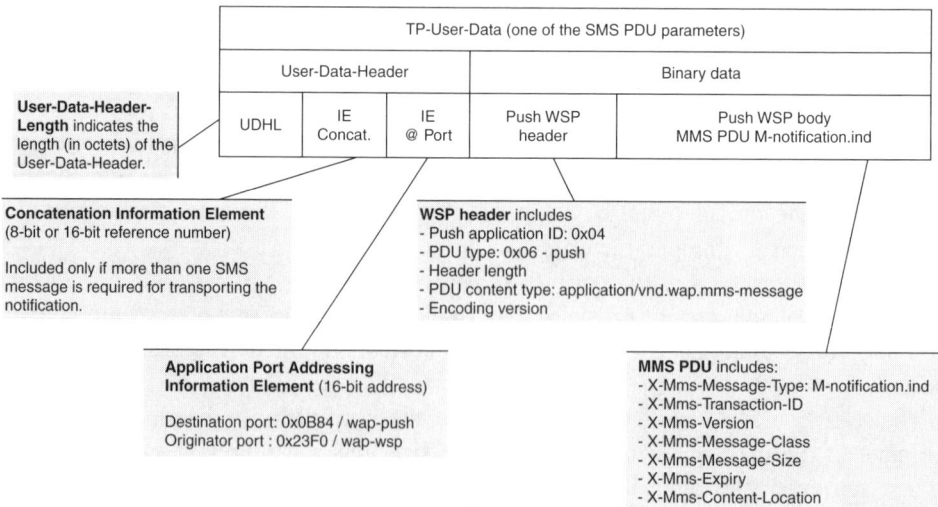

**Figure 6.11**   Structure of an MMS notification conveyed in a SMS message

WSP header

```
00000000h: 04 06 37 61 70 70 6C 69 63 61 74 69 6F 6E 2F 76  ; ..7application/v
00000010h: 6E 64 2E 77 61 70 2E 6D 6D 73 2D 6D 65 73 73 61  ; nd.wap.mms-messa
00000020h: 67 65 00 45 6E 63 6F 64 69 6E 67 2D 76 65 72 73  ; ge.Encoding-vers
00000030h: 69 6F 6E 00 31 2E 31 00 AF 84 8C 82 98 54 72 61  ; ion.1.1.»äîéÿTra
00000040h: 6E 73 61 63 74 69 6F 6E 2D 49 44 2D 31 32 33 34  ; nsaction-ID-1234
00000050h: 35 36 37 38 39 2D 61 62 63 64 65 66 67 00 8D 90  ; 56789-abcdefg.ìÉ
00000060h: 89 19 80 2B 33 33 36 36 36 36 36 36 36 36 36 36  ; ë.Ç+336666666666
00000070h: 2F 54 59 50 45 3D 50 4C 4D 4E 00 8A 80 8E 01 78  ; /TYPE=PLMN.èÇÄ.x
00000080h: 88 05 81 03 02 A3 00 83 68 74 74 70 3A 2F 2F 6D  ; ê.ü..ú.âhttp://m
00000090h: 6D 73 63 2E 70 72 6F 76 69 64 65 72 2D 64 6F 6D  ; msc.provider-dom
000000a0h: 61 69 6E 2E 63 6F 6D 3A 38 30 30 32 2F 70 61 74  ; ain.com:8002/pat
000000b0h: 68 2F 6D 65 73 73 61 67 65 2D 31 32 33 34 35 36  ; h/message-123456
```

MMS PDU

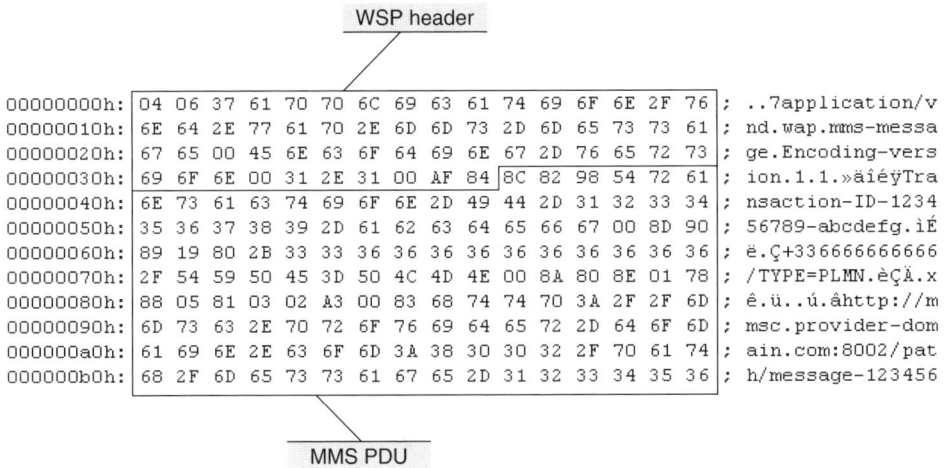

**Figure 6.12**   Hexadecimal trace dump for notification indication (M-notification.ind)

```
From: +336666666666/TYPE=PLMN
X-Mms-Message-Class: Personal
X-Mms-Message-Size: 120
X-Mms-Expiry: 172800
X-Mms-Content-Location:
   http://mmsc.provider-domain.com:8002/message-123456
```

The notification indication shown in Figure 6.12 is 192 bytes long and therefore fits into two SMS message segments (concatenated message). The User-Data-Header (without UDHL) of the first SMS message segment is shown in Table 6.5.

The User-Data-Header of the second SMS message segment is similar to that of the first SMS message segment. Only the segment index differs (segment index: 0x02). Consequently, the User-Data-Header (without UDHL) of the second SMS message is 0x000306020205040B8423F0.

The recipient MMS client confirms the reception of the notification indication PDU (M-notification.ind) with a notification response indication (M-notify-resp.ind). Parameters composing the notification response indication PDU are presented in Table 6.6. The notification response indication PDU does not contain any body (header only).

The parameter X-Mms-Status of the notification response indication informs on the status of the message. The following values can be assigned to this parameter:

- *Retrieved*: This status code indicates that the message has been successfully retrieved by the recipient MMS client (immediate retrieval).
- *Rejected*: This status code indicates that the recipient MMS client rejects the message (e.g. anonymous messages automatically rejected by the recipient MMS client).
- *Deferred*: This status code indicates that the message may be retrieved at a later stage (deferred retrieval).

**Table 6.5**  SMS User Data Header of the first SMS message segment

| [a]B.E. | | Description | | |
|---------|---|---|---|---|
| 0x00 | [b]IEI: | Concatenation 8-bit reference number (0x00) | | |
| 0x03 | [c]IEDL: | 3 Bytes (0x03) | | |
| 0x06 | [d]IED: | Byte 1: | Reference number : | 0x06 |
| 0x02 | | Byte 2: | Number of segments: | 0x02 |
| 0x01 | | Byte 3: | Segment index: | 0x01 |
| 0x05 | IEI: | Application port addressing scheme, 16-bit address (0x05) | | |
| 0x04 | IEDL: | 4 Bytes (0x04) | | |
| 0x0B | IED: | Bytes 1 . . . 2 | Destination port number: | 0x0B84 |
| 0x84 | | | | |
| 0x23 | | Bytes 3 . . . 4 | Source port number: | 0x23F0 |
| 0xF0 | | | | |

[a] B.E. stands for Binary Encoding.
[b] IEI stands for Information Element Identifier.
[c] IEDL stands for Information Element Data Length.
[d] IED stands for Information Element Data.

**Table 6.6**  MM1 notification response indication (`M-notifyresp.ind`)

| Parameter name | Description | From OMA | St. |
|---|---|---|---|
| X-Mms-Message-Type | MMS PDU type. Value: `M-notifyresp-ind` | 1.0 | ● |
| X-Mms-Transaction-ID | Unique identifier for the notification transaction. The same as the one for the corresponding notification indication. | 1.0 | ● |
| X-Mms-MMS-Version | MMS protocol version such as 1.0, 1.1 or 1.2. | 1.0 | ● |
| X-Mms-Status | Status of corresponding message such as 'retrieved', 'rejected', 'deferred' or 'unrecognised'. | 1.0 | ● |
| X-Mms-Report-Allowed | Indication whether or not the recipient MMS client allows the generation of a delivery report by the recipient MMSC. Possible values are 'yes' (default) or 'no'. | 1.0 | ○ |

- *Unrecognized*: This status code is for version management purpose only. For instance, this status code is used by MMS clients if they receive PDUs that are not recognized.

The initial case for element descriptors (parameter X-Mms-Element-Descriptor) was to describe, in the notification, media objects contained in a multimedia message to allow a selective retrieval of parts of multimedia messages. However, the latest version of the MMS standard (MMS 1.2, at the time of writing) does not provide support for such selective retrieval. It is therefore very unlikely that notifications will contain an element descriptor.

---

**BOX 6.1   Is the notification response indication required when notification is conveyed over the SMS bearer?**

It has been shown in this section that the MMS notification is conveyed over the SMS bearer or alternatively over a data connection (e.g. GSM data connection or GPRS) if one had previously been established by the MMS client (e.g. message submission or message retrieval). The successful retrieval of the message (immediate retrieval) or the receipt of a positive notification response (deferred retrieval) indicates to the MMSC that the notification indication has been successfully received by the MMS client. Assuming that the notification has been successfully received, the MMSC stops attempting to deliver the notification to the MMS client. Otherwise, the MMSC retransmits the notification after a period of time (period duration not defined in the MMS standards) until the notification is received by the MMS client or until the validity of the corresponding message expires. Theoretically, this behaviour should also apply in the situation in which the notification is conveyed over the SMS bearer. Practically, it happens that, in the deferred retrieval case, MMSCs stop attempting to deliver a notification over SMS as soon as the mobile device has successfully acknowledged SMS messages composing the notification at the SMS level, without expecting the MMS-level notification response indication. From an operator perspective, this bearer-dependent behaviour helps reduce signalling traffic for MMS and this behaviour avoids establishing a data connection (of particular interest in the roaming scenario when inter-operator contract agreements have not been set). Unfortunately, this deviation from the MMS standards is the subject of interoperability issues that can lead to message losses.

---

### 6.2.3 Message Retrieval

Message retrieval refers to the delivery of a multimedia message from the recipient MMSC to the recipient MMS client. A necessary condition for retrieving a message is that the corresponding notification has been received by the recipient MMS client. The transaction flow for message retrieval is shown in Figure 6.13. The PDU corresponding to the retrieval request is named `WSP/HTTP GET.req` and the corresponding confirmation is named `M-retrieve.conf`. If message retrieval is successful, then the retrieval confirmation PDU body contains the message. According to the

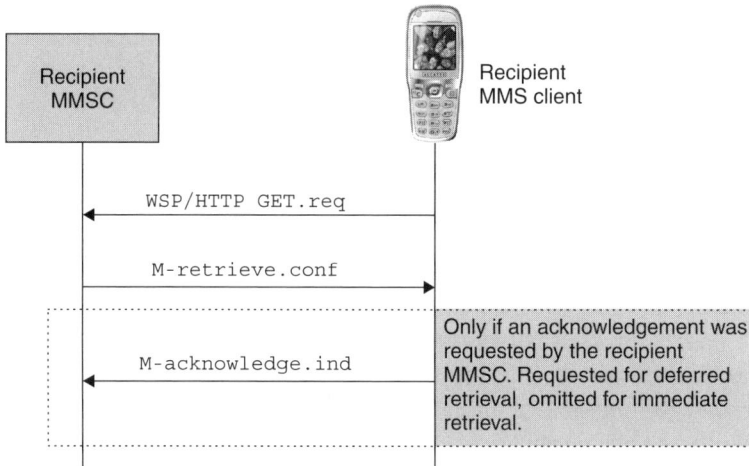

**Figure 6.13**   MM1 message retrieval transaction flow

retrieving mode, the MMS client may further acknowledge the message retrieval to the MMSC with a retrieval acknowledgement named `M-acknowledge.ind` (deferred retrieval only).

As shown in the previous chapter, a message can be retrieved according to two distinct modes: immediate and deferred retrievals.

With *immediate retrieval*, the recipient MMS client initiates the message retrieval automatically upon reception of the corresponding notification. The user is usually notified that a new message has been received once the message has been completely retrieved by the MMS client. In the immediate retrieval mode, the MMS client indicates to the MMSC that the notification has been successfully received (with the `M-notifyresp.ind` PDU) only after the corresponding message has been successfully retrieved (with the `M-retrieve.conf` PDU). With this method, the role of the `M-notifyresp.ind` PDU is twofold: first, it informs the MMSC that the notification has been successfully processed and second, that the corresponding message has been correctly retrieved. This explains why, in the immediate retrieval mode, the `M-acknowledge.ind` PDU (used to confirm the message retrieval) is not required. Owing to some transient error conditions, it may happen that message retrieval cannot be performed. In such a situation, the MMSC will retransmit the notification to the MMS client after a given timeout period (dependant on MMSC configuration).

With *deferred retrieval*, the MMS client does not retrieve the message automatically. Instead, the MMS client indicates to the recipient MMSC that the notification has been successfully received and informs the MMSC that the corresponding message may be retrieved at a later stage (e.g. upon user request). At this stage, the user is usually notified that a message awaits retrieval. It becomes the user's responsibility to initiate manually the retrieval of the message. With the deferred retrieval mode,

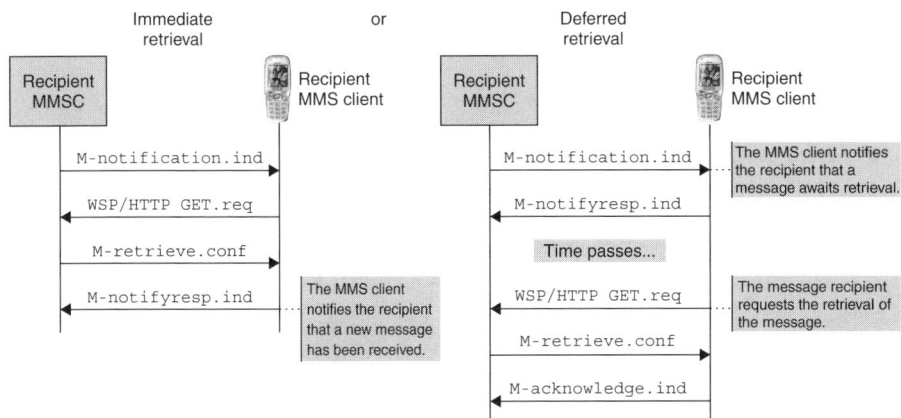

**Figure 6.14**   MM1 immediate and deferred retrieval transaction flows

the `M-acknowledge.ind` PDU is used by the MMS client to indicate to the MMSC that the message has been successfully retrieved.

The notification and message retrieval transaction flows are tightly interleaved and differ according to the selected retrieval mode. Figure 6.14 shows the two possible transaction flows.

No dedicated MMS PDU was designed for the message retrieval request. Instead, the existing object retrieval request from the WSP/HTTP protocol is used. This request is named `WSP/HTTP GET.req` and the object that is retrieved is the multimedia message. This message is found at the location provided in the corresponding notification request (`X-Mms-Content-Location` parameter of the `M-notification.ind` PDU). The message is delivered as part of the `M-retrieve.conf` PDU for which parameters are described in Table 6.7. The body of this PDU contains the message being retrieved or an error text if the message cannot be retrieved. Additionally, from MMS 1.1, if the message cannot be delivered by the MMSC, then the message retrieval confirmation PDU indicates an error code (parameter `X-Mms-Retrieve-Status`, see Appendix C). Once the message has been successfully retrieved, the MMS client acknowledges the successful retrieval with the `M-acknowledge.ind` PDU, only if requested by the MMSC (deferred retrieval only). The parameters of the retrieval acknowledgement PDU are described in Table 6.8.

From MMS 1.1, the MMSC may indicate the reason for retrieval failure by assigning appropriate values to the parameters `X-Mms-Retrieve-Status` and `X-Mms-Retrieve-Text`. Predefined status codes for the parameter `X-Mms-Retrieve-Status` are listed in Appendix C. The textual description assigned to the `X-Mms-Retrieve-Text` parameter can appropriately be based on code names defined in [RFC-1893]. In addition, the MMSC also provides an error text as part of the message body, when the `X-Mms-Retrieve-Status` parameter indicates an error, to ensure backward compatibility with MMS 1.0 clients.

**Table 6.7** MM1 message retrieval response (`M-retrieve.conf`)

| Parameter name | Description | From OMA | St. |
|---|---|---|---|
| X-Mms-Message-Type | MMS PDU type. Value: `M-retrieve-conf` | 1.0 | ● |
| X-Mms-Transaction-ID | Unique identifier for the retrieval transaction. This unique identifier is provided by the MMSC only if message retrieval is to be acknowledged with the `M-acknowledge.ind` PDU (deferred retrieval only). | 1.0 | ○ |
| X-Mms-MMS-Version | MMS protocol version such as 1.0, 1.1 or 1.2. | 1.0 | ● |
| Message-ID | Message unique identifier. This identifier is always provided if the PDU contains the message. It allows the MMS client to correlate read reports and reply messages (reply charging) with the original message. | 1.0 | ©[a] |
| Date | Date and time of latest submission or forwarding of the message (or reception by the MMSC). | 1.0 | ● |
| From | Address of the originator MMS client (phone number or email address). The 'insert token' is not allowed for this PDU. This parameter is not present in the message if the originator requested address hiding. | 1.0 | ○ |
| X-Mms-Previously-Sent-By | Address(es) of the MMS client(s) that have previously handled the message (submission or forward). This parameter may appear multiple times. | 1.1 | ○ |
| X-Mms-Previously-Sent-Date | Date and time when the message was previously handled by MMS clients (submission or forward). This parameter may appear multiple times. | 1.1 | ○ |
| To | One or multiple addresses (phone number or email address) for message recipient(s). Primary recipients. | 1.0 | ○ |
| Cc | One or multiple addresses (phone number or email address) for message recipient(s). Secondary recipients. | 1.0 | ○ |
| Subject | A short textual description for the message. | 1.0 | ○ |
| X-Mms-Message-Class | Message class such as 'auto' (automatically generated by the originator MMS client), 'personal' (default), 'advertisement' and 'informational'. Other classes can also be defined in the form of text strings. | 1.0 | ○ |
| X-Mms-Priority | Priority such as 'low', 'normal' (default) or 'high'. | 1.0 | ○ |
| X-Mms-Delivery-Report | Request for a delivery report. This parameter indicates whether or not delivery reports are to be generated for the message. Two values can be assigned to this parameter: 'yes' (delivery report is to be generated) or 'no' (no delivery report requested). | 1.0 | ○ |

**Table 6.7**  (*continued*)

| Parameter name | Description | From OMA | St. |
|---|---|---|---|
| X-Mms-Read-Report | Request for a read report. This parameter indicates whether or not a read report is to be generated for the message. Two values can be assigned to this parameter: 'yes' (read report is to be generated) or 'no' (no read report requested). | 1.0 | ○ |
| X-Mms-Reply-Charging | Request for reply charging. The presence of this parameter indicates that reply charging is requested by the message originator. Two values can be assigned to this parameter: 'requested' when the originator is willing to pay for the message reply(s) or 'requested text only' when the originator is willing to pay for message reply(s) containing text only. In any case, two parameters (reply message size and reply deadline) specify conditions for the message reply to be paid for by the originator. | 1.1 | ○ |
| X-Mms-Reply-Charging-Deadline | Reply charging – deadline. This parameter specifies the latest time for the recipient(s) to submit a message reply. This parameter is only present in the PDU if reply charging is requested. | 1.1 | ○ |
| X-Mms-Reply-Charging-Size | Reply charging – maximum message size. This parameter specifies the maximum size for message replies. This parameter is only present in the PDU if reply charging is requested. | 1.1 | ○ |
| X-Mms-Reply-Charging-ID | Reply charging – identifier. This parameter is inserted in a reply message only and refers to the original message identifier (Message-ID parameter). | 1.1 | ○ |
| X-Mms-Retrieve-Status | Status code for the retrieval transaction. The retrieval request can be accepted or rejected (permanent or transient errors). See error codes in Appendix C. | 1.1 | ○ |
| X-Mms-Retrieve-Text | Description of the transaction status. Descriptions can appropriately be based on the status code names from [RFC-1893]. | 1.1 | ○ |
| X-Mms-MM-State | MMBox message state. This parameter only appears for a message retrieved from an MMBox and indicates the message state (draft, sent, new, retrieved or forwarded). | 1.2 | ○ |
| X-Mms-MM-Flags | MMBox message flags. This parameter only appears for a message retrieved from an MMBox and indicates the lists of flags | 1.2 | ○ |

*(continued overleaf)*

**Table 6.7** (*continued*)

| Parameter name | Description | From OMA | St. |
|---|---|---|---|
| | associated to the message. The `X-Mms-Flags` parameter appears multiple times if the message is associated to several flags. | | |
| `X-Mms-Distribution-Indicator` | Message distribution indicator. This parameter can be present when the message is sent by a VAS provider. The value 'no' indicates to the recipient that the originator requested the content of the message not to be distributed further. | 1.2 | ○ |
| `Content-Type` | Content type of the multimedia message. | 1.0 | ● |

[a] In MMS 1.0, the `Message-ID` parameter is optional and not conditional.

**Table 6.8** MM1 message retrieval acknowledgement (`M-acknowledge.ind`)

| Parameter name | Description | From | St. |
|---|---|---|---|
| `X-Mms-Message-Type` | MMS PDU type. Value: `M-acknowledge-ind` | 1.0 | ● |
| `X-Mms-Transaction-ID` | Unique identifier for the retrieval transaction. The same as the one for the corresponding retrieval confirmation. | 1.0 | ● |
| `X-Mms-MMS-Version` | MMS protocol version such as 1.0, 1.1 or 1.2. | 1.0 | ● |
| `X-Mms-Report-Allowed` | Indication whether or not the recipient MMS client allows the generation of a delivery report by the recipient MMSC. Possible values are 'yes' (default) or 'no'. | 1.0 | ○ |

Note that for the retrieval acknowledgement PDU, the transaction identifier is the same as the one provided by the recipient MMSC for the corresponding retrieval confirmation PDU (`M-retrieve.conf` PDU). Depending on MMSC implementations, the transaction identifier may also be the same as the one provided for the corresponding notification indication PDU (`M-notification.ind` PDU).

The MMSC may indicate that a secure transfer protocol is to be used by the MMS client for the retrieval of the message by indicating so in the corresponding message notification. For instance, this can be done by using the protocol scheme `https` as part of the `X-Mms-Content-Location` parameter of the corresponding notification (e.g. `https//mmsc.operator.net/msgid-6543210`). See also Section 4.14.

### 6.2.4 Delivery Report

During message submission, the originator MMS client has the possibility of requesting the generation of delivery report(s) for the submitted message. If such a request has been made, a delivery report may be generated for each one of the message recipients on the occurrence of the following events:

- Message has been successfully retrieved by the recipient MMS client.
- Message has been deleted (e.g. validity period has expired).
- Message has been rejected by the recipient MMS client.
- Message has been forwarded by the recipient MMS client.
- Message status is indeterminate (e.g. message has been transferred to a messaging system where the concept of delivery report is not fully supported).

Note that a recipient MMS client can deny the generation of delivery reports (parameter X-Mms-Report-Allowed for M-acknowledge.ind and M-notifyresp.ind PDUs). If allowed by the recipient MMS client, the recipient MMSC generates the delivery report and forwards it to the originator MMSC over the MM4 interface. Upon receipt of the delivery report, the originator MMSC delivers it to the originator MMS client over the MM1 interface with the M-delivery.ind PDU as shown in Figure 6.15.

Parameters of the M-delivery.ind PDU are described in Table 6.9.

Originator MMS client

Originator MMSC

M-delivery.ind

**Figure 6.15** MM1 delivery report transaction flow

**Table 6.9** MM1 delivery report indication (M-delivery.ind)

| Parameter name | Description | From OMA | St. |
|---|---|---|---|
| X-Mms-Message-Type | MMS PDU type. Value: M-delivery-ind | 1.0 | ● |
| X-Mms-MMS-Version | MMS protocol version such as 1.0, 1.1 or 1.2. | 1.0 | ● |
| Message-ID | Identifier of the message to which the report relates. This identifier is used for correlating the delivery report with the original message. | 1.0 | ● |
| To | Address of the message recipient. | 1.0 | ● |
| Date | Date and time the message was retrieved, has expired, etc. | 1.0 | ● |
| X-Mms-Status | Status of corresponding message such as 'expired', 'retrieved', 'rejected', 'forwarded' or 'indeterminate'. | 1.0 | ● |

```
                              ┌──────────────┐
                              │  WSP header  │
                              └──────────────┘
                                    ╱
00000000h: │ 04 06 37 61 70 70 6C 69 63 61 74 69 6F 6E 2F 76 │ ; ..7application/v
00000010h: │ 6E 64 2E 77 61 70 2E 6D 6D 73 2D 6D 65 73 73 61 │ ; nd.wap.mms-messa
00000020h: │ 67 65 00 45 6E 63 6F 64 69 6E 67 2D 76 65 72 73 │ ; ge.Encoding-vers
00000030h: │ 69 6F 6E 00 31 2E 31 00 AF 84 │8C 86 8D 90 8B 4D │ ; ion.1.1.»âîâÎÉÎM
00000040h: │ 65 73 73 61 67 65 2D 31 32 33 34 35 00 97 2B 33 │ ; essage-12345.ù+3
00000050h: │ 33 36 36 36 36 36 36 36 36 36 2F 54 59 50 45 2B │ ; 3666666666/TYPE+
00000060h: │ 50 4C 4D 4E 00 85 04 35 3F 45 11 95 81          │ ; PLMN.à.5?E.òü
                              ┌──────────────┐
                              │   MMS PDU    │
                              └──────────────┘
```

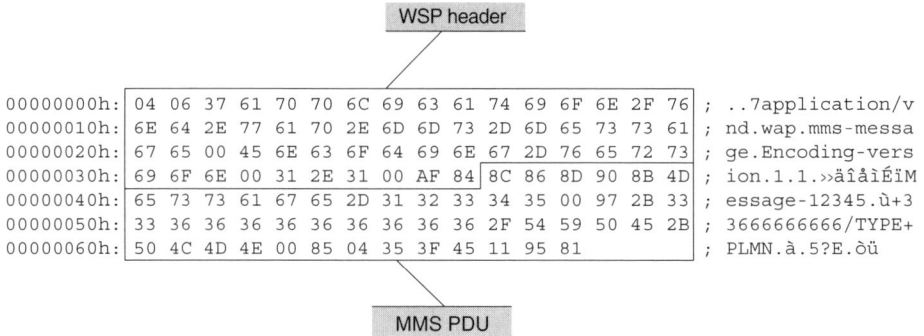

**Figure 6.16**   Hexadecimal trace dump for delivery report

**Table 6.10**   SMS User-Data-Header of the first message segment

| [a]B.E. | | | Description | |
|---------|---|---|-------------|---|
| 0x05 | [b]IEI: | | Application port addressing scheme, 16-bit address (0x05) | |
| 0x04 | [c]IEDL: | | 4 Bytes (0x04) | |
| 0x0B | [d]IED: | | Bytes 1...2 | Destination port number:  0x0B84 |
| 0x84 | | | | |
| 0x23 | | | Bytes 3...4 | Source port number:      0x23F0 |
| 0xF0 | | | | |

[a] B.E. stands for Binary Encoding.
[b] IEI stands for Information Element Identifier.
[c] IEDL stands for Information Element Data Length.
[d] IED stands for Information Element Data.

Delivery and read reports do not contain a transaction identifier since reports are never acknowledged over the MM1 interface.

Figure 6.16 shows a hexadecimal trace dump for a delivery report with the following characteristics:

```
X-Mms-Message-Type: M-delivery.ind
X-Mms-MMS-Version: 1.0
Message-ID: Message-12345
To: +33666666666/TYPE=PLMN
Date: Thu, 23 Apr 1998 13:41:37 GMT
X-Mms-Status: Retrieved
```

The delivery report indication shown in Figure 6.16 is 109 bytes long and therefore fits into one single SMS message (no need for SMS message concatenation). The User-Data-Header (without UDHL) of the SMS message is shown in Table 6.10.

### 6.2.5 Read Report

During message submission, the originator MMS client has the possibility of requesting the generation of read report(s) for the submitted message. If such a request has been made, a read report may be generated for each one of the message recipients on the occurrence of the following events:

- Message has been read.
- Message has been deleted without being read.

Unlike delivery reports, it is usually the responsibility of the recipient MMS client to generate the read report when the message has been read by the recipient or deleted without being read. The recipient may deny the generation of a read report (MMS client setting/according to device features). The MMS client can generate the read report according to two different methods:

- The first method, introduced in MMS 1.0, lets the MMS client submit a normal message addressed to the message originator with the `M-send.req` PDU over the MM1 interface. In this case, the message class is set to 'auto' and requests for read and delivery reports are disabled. With this method, it is up to the recipient MMS client to include appropriate message content in the message submission PDU body indicating the status of the corresponding message (along with original message subject and message identifier). This first method does not allow the originator MMS client to easily match the received read report with the corresponding submitted message.
- The second method, applicable from MMS 1.1, consists in using a set of two MM1 PDUs dedicated to read reports as shown in Figure 6.17. This method relies on three successive steps:
  1. The recipient MMS client provides the read report to the recipient MMSC over the MM1 interface as part of an `M-read-rec.ind` PDU.
  2. Upon receipt of the read report, the recipient MMSC routes forward the read report to the originator MMSC over the MM4 interface.
  3. The originator MMSC provides the read report to the originator MMS client over the MM1 interface as part of an `M-read-orig.ind` PDU.

Parameters of the `M-read-rec.ind` and `M-read-orig.ind` PDUs are described in Table 6.11.

If the optional `Date` parameter is not present in the `M-read-rec.ind` PDU, then it becomes the responsibility of the recipient MMSC to update the read report with the appropriate date.

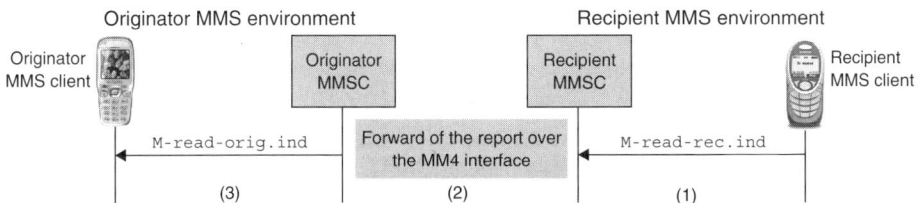

**Figure 6.17**  MM1 read-report transaction flow

**Table 6.11** MM1 read-report indications (`M-read-rec.ind` and `M-read-orig.ind`)

| Parameter name | Description | From OMA | St. |
|---|---|---|---|
| `X-Mms-Message-Type` | MMS PDU type. Value: `M-read-rec-ind` or `M-read-orig-ind` | 1.1 | ● |
| `X-Mms-MMS-Version` | MMS protocol version such as 1.1 or 1.2. These PDUs are not defined for MMS 1.0. | 1.1 | ● |
| `Message-ID` | Identifier of the message to which the report relates. This identifier is used for correlating the read report with the original message. | 1.1 | ● |
| `To` | Address of the recipient of the read report (message originator). | 1.1 | ● |
| `From` | Address of the sender of the read report (message recipient). For the `M-read-rec.ind` PDU, an 'insert-token' can be used if the sender address is to be provided by the recipient MMSC. | 1.1 | ● |
| `Date` | Date and time the message was read or was deleted without being read. This parameter is optional for the `M-read-rec.ind` PDU and mandatory for the `M-read-orig.ind` PDU. | 1.1 | ○● |
| `X-Mms-Read-Status` | Status of corresponding message such as 'read' or 'deleted without being read'. | 1.1 | ● |

There are potential interoperability issues in an environment where communicating MMS entities comply with different methods for the management of read reports. The main risk is for the configuration in which the recipient MMS client and recipient/originator MMSCs comply with MMS 1.1 (or later) and the originator MMSC only complies with MMS 1.0. In this configuration, the originator MMS client may receive a dedicated read-report PDU that it cannot understand. The originator MMSC plays an important role in solving this interoperability issue. When the originator MMSC (complying with MMS 1.1 or later) becomes aware that the originator MMS client complies with MMS 1.0 only, then it can appropriately convert the dedicated read report into a normal message. The MMSC can identify that the originating MMS client complies with MMS 1.0 only, by means of the User Agent Profile (UAProf) mechanism (MMS version parameter, as described in Section 4.11). Alternatively, the MMS client generating an `M-notifyresp.ind` PDU containing a status value set to 'Unrecognised' upon receipt of dedicated read report (see Section 7.2) means that the MMS client complies with MMS 1.0 only.

## 6.2.6 Message Forward

Once a notification has been received by the recipient MMS client, the user has the possibility to retrieve or reject the corresponding message. In addition, from MMS 1.1, the recipient MMS client may also support the forward of a message that has not yet been retrieved from the MMSC. For instance, a user may wish to forward a large message to an email server for later viewing from a personal computer. The transaction flow between the MMS client and the recipient MMSC is shown in Figure 6.18. The PDU corresponding to the forward request is named `M-forward.req` and the corresponding confirmation is named `M-forward.conf`.

Parameters of the `M-forward.req` PDU are described in Table 6.12, whereas parameters of the `M-forward.conf` PDU are described in Table 6.13.

Recipient addressing parameters (`To`, `Cc` and `Bcc`) are all optional in the forward request. However, at least one recipient address shall be provided by the forwarding MMS client to the MMSC.

If the concept of MMBox is not supported by the originator MMSC, then the forward request MMBox-related parameters (`X-Mms-Store`, `X-Mms-MM-State` and `X-Mms-MM-Flags`) are ignored by the MMSC.

When accepting the forwarding request, the MMSC assigns the address of the forwarding MMS client to the `From` parameter of the forwarded message. Optionally, the MMSC assigns the previous address assigned to the `From` parameter to a new instance of the `X-Mms-Previously-Sent-By` parameter and associates a sequence number to this parameter. This sequence number is an increment of the highest sequence number of `X-Mms-Previously-Sent-By` parameters already present in the forwarded message. If there is no `X-Mms-Previously-Sent-By` parameter present in the forwarded message, then the new `X-Mms-Previously-Sent-By` parameter is associated to the sequence number '0' (identifying the original sender of the message). The example below shows a list of `X-Mms-Previously-Sent-By` parameters identifying the original sender of the message (number '0', `gwenael@lebodic.net`) and users who forwarded the message (numbers '1' and '2'):

```
X-Mms-Previously-Sent-By: 0, Gwenael <gwenael@lebodic.net>
X-Mms-Previously-Sent-By: 1, +33612345678/TYPE=PLMN
X-Mms-Previously-Sent-By: 2, +33698765432/TYPE=PLMN
```

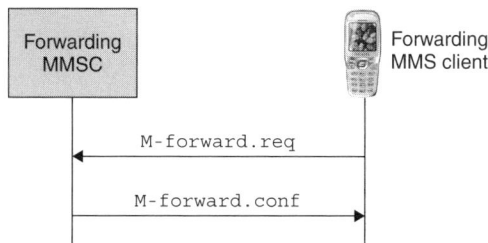

**Figure 6.18**   MM1 forward transaction flow

**Table 6.12**  MM1 forward request (`M-forward.req`)

| Parameter name | Description | From OMA | St. |
|---|---|---|---|
| X-Mms-Message-Type | MMS PDU type. Value: `M-forward-req` | 1.1 | ● |
| X-Mms-Transaction-ID | Unique identifier for the forward transaction. | 1.1 | ● |
| X-Mms-MMS-Version | MMS protocol version such as 1.1 or 1.2. This PDU is not defined for MMS 1.0. | 1.1 | ● |
| Date | Date and time of the forward transaction. | 1.1 | ○ |
| From | Address of the forwarding MMS client (phone number or email address) or 'insert token' if address is to be provided by the MMSC. | 1.1 | ● |
| To | One or multiple addresses (phone number or email address) for recipient(s) of the forwarded message. Primary recipients. | 1.1 | ○ |
| Cc | One or multiple addresses (phone number or email address) for recipient(s) of the forwarded message. Secondary recipients. | 1.1 | ○ |
| Bcc | One or multiple addresses (phone number or email address) for recipient(s) of the forwarded message. Secondary recipients/blind copy. | 1.1 | ○ |
| X-Mms-Expiry | Expiry date. Default value for this parameter is 'maximum'. | 1.1 | ○ |
| X-Mms-Delivery-Time | Earliest delivery time. Default value for this parameter is 'immediate delivery'. | 1.1 | ○ |
| X-Mms-Report-Allowed | Indication whether or not the forwarding MMS client allows the generation of a delivery report by the forwarding MMSC. Possible values are 'yes' (default) or 'no'. | 1.1 | ○ |
| X-Mms-Delivery-Report | Request for a delivery report. This parameter indicates whether or not delivery reports are to be generated for the forwarded message. Two values can be assigned to this parameter: 'yes' (delivery report is requested) or 'no' (no delivery report requested). | 1.1 | ○ |

**Table 6.12**   (*continued*)

| Parameter name | Description | From OMA | St. |
|---|---|---|---|
| X-Mms-Read-Report | Request for a read report. This parameter indicates whether or not read reports are to be generated for the forwarded message. Two values can be assigned to this parameter: 'yes' (read report is requested) or 'no' (no read report requested). | 1.1 | O |
| X-Mms-Store | MMBox storage request. This parameter indicates whether the forwarding MMS client requests to save the message in the user's MMBox in addition to sending it. | 1.2 | O |
| X-Mms-MM-State | MMBox message state. If X-Mms-Store is set to 'yes' or when X-Mms-Store is not present and X-Mms-Content-Location refers to a message stored in the MMBox then this parameter indicates the message state to be set in the originator's MMBox. When X-Mms-Store is set to 'yes' and if this parameter is not present then the message default state in the MMBox is set to 'forwarded'. | 1.2 | O |
| X-Mms-MM-Flags | MMBox message flags. This parameter indicates the list of flags associated to a message stored in the MMBox (considered only if X-Mms-Store is set to 'yes'). | 1.2 | O |
| X-Mms-Content-Location | Location of the message to be forwarded (as specified in the corresponding notification). | 1.1 | ● |

Additionally, the MMSC can insert the date and time contained in the Date parameter of the message to be forwarded into a new parameter X-Mms-Previously-Sent-Date and assign a sequence number to that parameter. In that case, the sequence number is the same as the sequence number of the corresponding X-Mms-Previously-Sent-By header field. The MMSC provides a corresponding X-Mms-Previously-Sent-By parameter for each X-Mms-Previously-Sent-Date parameter.

**Table 6.13**  MM1 message forward confirmation (`M-forward.conf`)

| Parameter name | Description | From OMA | St. |
|---|---|---|---|
| X-Mms-Message-Type | MMS PDU type. Value: M-forward-conf | 1.1 | ● |
| X-Mms-Transaction-ID | Unique identifier for the forward transaction. The same as the one for the corresponding forward request. | 1.1 | ● |
| X-Mms-MMS-Version | MMS protocol version such as 1.1 or 1.2. | 1.1 | ● |
| X-Mms-Response-Status | Status code for the forward transaction. The request can be accepted or rejected (permanent or transient errors). See status codes in Appendix B. | 1.1 | ● |
| X-Mms-Response-Text | Human readable description of the transaction status. | 1.1 | ○ |
| Message-ID | Message unique identifier. This identifier is always provided by the MMSC if the forward request is accepted. | 1.1 | ○ |
| X-Mms-Content-Location | Reference to the message stored in the MMBox. This parameter is present only if the three following conditions are fulfilled: <br> – the originator MMSC supports the MMBox feature <br> – the X-Mms-Store parameter was present in the corresponding forward request <br> – the X-Mms-Store-Status indicates 'success' <br> When available, this parameter provides a reference to the message stored in the MMBox (reference used later for message retrieval or view request). | 1.2 | ○ |
| X-Mms-Store-Status | MMBox store request status – This parameter is present only if the two following conditions are fulfilled: <br> – the originator MMSC supports the MMBox feature <br> – the X-Mms-Store parameter was present in the corresponding forward request | 1.2 | ○ |

**Table 6.13**  (*continued*)

| Parameter name | Description | From OMA | St. |
|---|---|---|---|
| | When available, this parameter indicates whether or not the forwarded message has been successfully stored in the MMBox. See status codes in Appendix D. | | |
| X-Mms-Store-Status-Text | MMBox message textual status. Textual description qualifying the value assigned to the X-Mms-Store-Status parameter. | 1.2 | O |

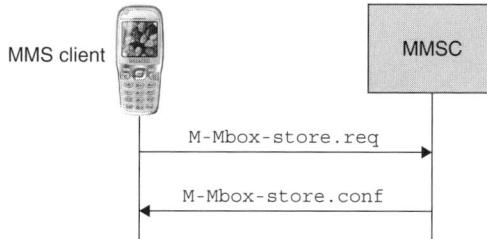

**Figure 6.19**  MM1 MMBox store/update transaction flow

The MMSC may forbid the forward of a message containing protected media objects (see Section 5.6).

### 6.2.7 Storing and Updating a Message in the MMBox

From MMS 1.2, the MMS client has the possibility to store or update a message in an MMBox (only if the concept of MMBox is supported by the MMSC):

- *Store message in MMBox*:  The MMS client instructs the MMSC to store a message, which has yet to be retrieved, in the MMBox.
- *Update message in MMBox*:  The MMS client instructs the MMSC to update the state and/or flags of a message in the MMBox.

The transaction flow for storing or updating a message in the MMBox is shown in Figure 6.19. The storing/updating request is named M-Mbox-store.req and is acknowledged with the M-Mbox-store.conf confirmation.

Parameters of the M-Mbox-store.req PDU are described in Table 6.14 and parameters of the M-Mbox-store.conf PDU are described in Table 6.15.

**Table 6.14**   MM1 MMBox store/update request (`M-Mbox-store.req`)

| Parameter name | Description | From OMA | St. |
|---|---|---|---|
| X-Mms-Message-Type | MMS PDU type. Value: `M-Mbox-store-req` | 1.2 | ● |
| X-Mms-Transaction-ID | Unique identifier for the store/update transaction. | 1.2 | ● |
| X-Mms-MMS-Version | MMS protocol version such as 1.2. | 1.2 | ● |
| X-Mms-Content-Location | Reference of the message to be stored or updated in the MMBox. | 1.2 | ● |
| X-Mms-MM-State | MMBox message state. If the `X-Mms-Content-Location` parameter refers to a message to be stored (reference retrieved from the `M-notification.ind` PDU), then the default state for the message state is 'new'. | 1.2 | ○ |
| X-Mms-MM-Flags | MMBox message flags. This parameter indicates the list of flags associated to a message stored in the MMBox (considered only if `X-Mms-Store` is set to 'yes'). This field may appear multiple times in the PDU. | 1.2 | ○ |

**Table 6.15**   MM1 MMBox store/update confirmation (`M-Mbox-store.conf`)

| Parameter name | Description | From OMA | St. |
|---|---|---|---|
| X-Mms-Message-Type | MMS PDU type. Value: `M-Mbox-store-conf` | 1.2 | ● |
| X-Mms-Transaction-ID | Unique identifier for the store/update transaction. The same as the one for the corresponding MMBox store/update request. | 1.2 | ● |
| X-Mms-MMS-Version | MMS protocol version such as 1.2. | 1.2 | ● |
| X-Mms-Content-Location | This parameter can be present only if the `X-Mms-Store-Status` parameter is set to 'success'. When available, this parameter indicates a reference to the message in the MMBox. This reference can be used subsequently for retrieving the message or for obtaining information about the message. | 1.2 | ○ |
| X-Mms-Store-Status | MMBox store request status. The status of the store/update request such as 'success'. See status codes in Appendix D. | 1.2 | ● |
| X-Mms-Store-Status-Text | MMBox message textual status. Textual description qualifying the value assigned to the `X-Mms-Store-Status` parameter. | 1.2 | ○ |

The MMSC determines the nature (store or update) of the request by analyzing the X-Mms-Content-Location parameter of the request:

- If the X-Mms-Content-Location refers to a message that is not yet in the user's MMBox, then the MMSC moves the message to the MMBox. The message is associated with the state and flags optionally set in the request. If no state was specified, then the message state is set to 'new' by default.
- If the X-Mms-Content-Location refers to a message already stored in the user's MMBox, then the MMSC updates the message with the state and flags specified as part of the request.

If the MMSC has successfully processed the request, then it provides a valid reference as part of the X-Mms-Content-Location parameter of the confirmation. This reference can be used subsequently for retrieving the message or for obtaining information related to the message.

### 6.2.8 Viewing Information from the MMBox

From MMS 1.2, the MMS client has the possibility to request information about one or more multimedia messages that are stored in the user's MMBox. The information not only includes message parameters (e.g. class, priority, originator address, etc.) but also the message contents if requested (e.g. media objects and scene description). The transaction flow for requesting such information is shown in Figure 6.20. The information-viewing request is named M-Mbox-view.req and is acknowledged with the M-Mbox-view.conf confirmation.

The identification of messages in the MMBox is performed according to one of the two following methods:

- Identifying the message with its Uniform Resource Locator (URL) assigned to the parameter X-Mms-Content-Location (obtained from a previous notification or store transaction).
- Filtering messages with criteria based on the message MMBox flags and state.

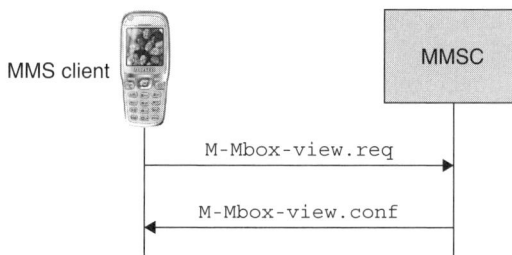

**Figure 6.20**   MM1 MMBox view transaction flow

For efficiency, the MMS client can retrieve information for all messages meeting the filtering criteria with several viewing transactions. This helps in reducing the amount of information contained in each single viewing transaction confirmation. For this purpose, the MMS client specifies the subset of messages (start and end of range) to be selected for each single viewing transaction. For instance, if 20 messages meet the filtering criteria, then the MMS client would be able to first retrieve the first 5 messages in one view transaction, the next 5 messages in a second view transaction, and so on. In addition, the MMS client indicates, as part of the viewing request, which message parameters/contents are to be returned as part of the viewing confirmation.

Upon reception of the viewing request, the MMSC selects the messages matching the selection criteria in the MMBox. The next step for the MMSC consists of generating the viewing confirmation as a multipart mixed structure for which each body part contains the requested information for one of the selected messages. Additionally, the MMSC provides the MMBox total and quota information as part of the viewing confirmation when requested by the MMS client.

Parameters of the `M-Mbox-view.req` PDU are described in Table 6.16 and parameters of the `M-Mbox-view.conf` PDU are described in Table 6.17. The view request PDU does not contain any body (header only).

**Table 6.16** MM1 MMBox view request (`M-Mbox-view.req`)

| Parameter name | Description | From OMA | St. |
|---|---|---|---|
| X-Mms-Message-Type | MMS PDU type. Value: `M-Mbox-view-req` | 1.2 | ● |
| X-Mms-Transaction-ID | Unique identifier for the view transaction. | 1.2 | ● |
| X-Mms-MMS-Version | MMS protocol version such as 1.2. | 1.2 | ● |
| X-Mms-Content-Location | Reference(s) of the message in the MMBox for which information is requested. This parameter may appear multiple times for obtaining information for several messages. | 1.2 | O |
| X-Mms-MM-State | MMBox message state. This parameter specifies the state of the message(s) for which information is requested. When this parameter appears multiple times, the selection is the union of all messages fulfilling at least one of the parameter requirements (e.g. if information is requested for draft and sent messages, then this parameter appears twice). | 1.2 | O |
| X-Mms-MM-Flags | MMBox message flags. This parameter specifies the flag of the message(s) for which information is requested. When this parameter appears multiple times, then the selection is the union of all messages fulfilling at least one of the parameter requirements. | 1.2 | O |

**Table 6.16** (*continued*)

| Parameter name | Description | From OMA | St. |
|---|---|---|---|
| X-Mms-Start | Start of selection range. A number indicating the first multimedia message in the selection for which information is returned in the transaction confirmation. | 1.2 | ○ |
| X-Mms-Limit | Size of selection range. The number of multimedia messages in the selection for which information is returned in the transaction confirmation. | 1.2 | ○ |
| X-Mms-Attributes | Message parameter(s) to be returned. This parameter indicates the list of message parameters (message information) to be included as part of the transaction confirmation. If not present, then an MMSC default set of message parameters is considered. If no default set has been defined, then the set of message parameters composing the notification is used instead. | 1.2 | ○ |
| X-Mms-Totals | Request for returning MMBox totals. This parameter indicates whether or not the MMS client requests the total count of messages stored in the user's MMBox to be returned as part of the transaction confirmation. Two values can be assigned to this parameter: 'yes' (information on totals is requested) or 'no' (no information on totals requested). If not present, then the default value is 'no' request. | 1.2 | ○ |
| X-Mms-Quotas | Request for returning MMBox quotas. This parameter indicates whether or not the MMS client requests the user's MMBox quotas to be returned as part of the transaction confirmation. Two values can be assigned to this parameter: 'yes' (information on quotas is requested) or 'no' (no information on quotas requested). If not present, then the default value is 'no' request. | 1.2 | ○ |

**Table 6.17** MM1 MMBox view confirmation (M-Mbox-view.conf)

| Parameter name | Description | From OMA | St. |
|---|---|---|---|
| X-Mms-Message-Type | MMS PDU type. Value: M-Mbox-view-conf | 1.2 | ● |
| X-Mms-Transaction-ID | Unique identifier for the view transaction. The same as the one for the corresponding MMBox view request. | 1.2 | ● |

(*continued overleaf*)

**Table 6.17**   (*continued*)

| Parameter name | Description | From OMA | St. |
|---|---|---|---|
| X-Mms-MMS-Version | MMS protocol version such as 1.2. | 1.2 | ● |
| X-Mms-Response-Status | Status code for the view transaction. The view request can be accepted or rejected (permanent or transient errors). See status codes in Appendix B. | 1.2 | ● |
| X-Mms-Response-Text | Human readable description of the transaction status. | 1.2 | ○ |
| X-Mms-Content-Location | Copied from the corresponding request – Reference(s) of the message in the MMBox for which information is requested. This parameter may appear multiple times for obtaining information for several messages. This parameter appears in the confirmation only if it is present in the corresponding request. | 1.2 | ○ |
| X-Mms-MM-State | Copied from the corresponding request – State criteria for selected messages. This parameter may appear several times and appears in the confirmation only if it is present in the corresponding request. | 1.2 | ○ |
| X-Mms-MM-Flags | Copied from the corresponding request – Flag criteria for selected messages. This parameter may appear several times and appears in the confirmation only if it is present in the corresponding request. | 1.2 | ○ |
| X-Mms-Start | Copied from the corresponding request – A number indicating the first multimedia message in the selection for which information is returned. This parameter appears in the confirmation only if it is present in the corresponding request. | 1.2 | ○ |
| X-Mms-Limit | Copied from the corresponding request – The number of multimedia messages in the selection for which information is returned. This parameter appears in the confirmation only if it is present in the corresponding request. | 1.2 | ○ |
| X-Mms-Attributes | Copied from the corresponding request or specified by the MMSC. This parameter indicates the list of message parameters (message information) included as part of the transaction confirmation. | 1.2 | ○ |
| X-Mms-Mbox-Totals | MMBox totals. This parameter indicates the total number of messages or bytes in the user's MMBox. This parameter appears in the confirmation if the parameter X-Mms-Totals | 1.2 | ○ |

**Table 6.17**  (*continued*)

| Parameter name | Description | From OMA | St. |
|---|---|---|---|
| | of the corresponding request is set to 'yes'. Otherwise, it does not appear in the confirmation. | | |
| X-Mms-Mbox-Quotas | MMBox quotas. This parameter indicates the quotas of the user's MMBox in terms of messages or bytes. This parameter appears in the confirmation if the parameter X-Mms-Quotas of the corresponding request is set to 'yes'. Otherwise, it does not appear in the confirmation. | 1.2 | ○ |
| X-Mms-Message-Count | Number of messages for which information is included in the confirmation. | 1.2 | ○ |
| Content-type | This parameter always appears as the last parameter of the PDU header. If the PDU has no body then the content type '*/*' is used (binary encoded as 0x00). Otherwise, the content type is 'application/vnd.multipart.mixed'. | 1.2 | ● |

**Table 6.18**  Parameters of the structure containing information related to a viewed message

| Parameter name | Description | From OMA | St. |
|---|---|---|---|
| X-Mms-Message-Type | MMS PDU type. Value: M-Mbox-descr | 1.2 | ● |
| X-Mms-Content-Location | Location of the message. This reference can be used by the MMS client for subsequent requests related to the corresponding message. This parameter is only used in conjunction with a view transaction. | 1.2 | © |
| Message-ID | Message unique identifier. | 1.2 | ● |
| X-Mms-MM-State | MMBox message state. | 1.2 | ● |
| X-Mms-MM-Flags | MMBox message flags. This parameter may appear multiple times. | 1.2 | ○ |
| Date | Date and time of message submission. Note (A). | 1.2 | ○ |
| From | Address of the originator MMS client. Note (A). | 1.2 | ○ |
| To | One or multiple addresses (phone number or email address) for message recipient(s). Primary recipients. Note (A). | 1.2 | ○ |
| Cc | One or multiple addresses (phone number or email address) for message recipient(s). Secondary recipients. Note (A). | 1.2 | ○ |

(*continued overleaf*)

**Table 6.18**   (*continued*)

| Parameter name | Description | From OMA | St. |
|---|---|---|---|
| Bcc | One or multiple addresses (phone number or email address) for message recipient(s). Secondary recipients/blind copy. Note (A). | 1.2 | O |
| X-Mms-Message-Class | Message class. Note (A). | 1.2 | O |
| Subject | A short textual description for the message. Note (A). | 1.2 | O |
| X-Mms-Priority | Priority such as 'low', 'normal' (default) or 'high'. Note (A). | 1.2 | O |
| X-Mms-Delivery-Time | Earliest delivery time. Note (A). | 1.2 | O |
| X-Mms-Expiry | Expiry time. Note (A). | 1.2 | O |
| X-Mms-Delivery-Report | Request for a delivery report. Note (A). | 1.2 | O |
| X-Mms-Read-Report | Request for a read report. Note (A). | 1.2 | O |
| X-Mms-Message-Size | Approximate size of the message. | 1.2 | O |
| X-Mms-Reply-Charging | Request for reply charging. Note (A). | 1.2 | O |
| X-Mms-Reply-Charging-ID | Reply charging – identification. This parameter is inserted in a reply message only and refers to the original message identification (Message-ID parameter). | 1.2 | O |
| X-Mms-Reply-Charging-Deadline | Reply charging – deadline. Note (A). | 1.2 | O |
| X-Mms-Reply-Charging-Size | Reply charging – maximum message size. Note (A). | 1.2 | O |
| X-Mms-Previously-Sent-By | Address(es) of the MMS client(s) that have previously handled the message (submission or forward). Note (A). | 1.2 | O |
| X-Mms-Previously-Sent-Date | Date and time when the message was previously handled by MMS clients (submission or forward). Note (A). | 1.2 | O |
| Content-Type | Content type of the multimedia message. Note (A). | 1.2 | O |

*Note*: (A) In conjunction with a view transaction, the parameter appears only if requested by the MMS client (according to parameter X-Mms-Attributes of the corresponding view request).

If the MMS client view request can be processed successfully, then the body of the confirmation PDU is organized as a multipart structure. For this purpose, the confirmation PDU content type is set to 'application/vnd.multipart.mixed'. Each body part of the structure contains information related to one of the selected messages as shown in Table 6.18.

Figure 6.21 shows how a view confirmation can contain information regarding two selected messages (textual representation).

**WSP/HTTP header**
At the transport level, MMS PDUs are identified by the content type 'application/vnd.mms-message'

```
...
Content-Type:application/vnd.wap.mms-message
...
```

**MMS PDU header**
This header of the view confirmation includes the message selection criteria.

```
X-Mms-Message-Type: M-Mbox-view-conf
X-Mms-Transaction-ID: 0123456789
X-Mms-MMS-Version: 1.2
X-Mms-Response-Status: Ok
X-Mms-MM-State: Draft
X-Mms-MM-State: New
X-Mms-Attributes: Subject
X-Mms-Attributes: From
Content-type:application/vnd.wap.multipart.mixed
                          boundary=boundary1
```

```
--boundary1
Content-type: application/vnd.wap.mms-message
X-Mms-Message-Type: m-mbox-view-descr
X-Mms-Content-Location: http://mmsc.green.net/mmbox/m336
Message-ID: m336-1299856
X-Mms-MM-State: Draft
Subject: A week-end in Velen
From: gwenael@lebodic.net
```
Information for the 1st message.

**MMS PDU body**
The body contains information related to selected messages.

```
--boundary1
Content-type: application/vnd.wap.mms-message
X-Mms-Message-Type: m-mbox-view-descr
X-Mms-Content-Location: http://mmsc.green.net/mmbox/m399
Message-ID: m399-45875241215
X-Mms-MM-State: New
Subject: Idea for next week
From: marie@lebodic.net

--boundary1--
```
Information for the 2nd message.

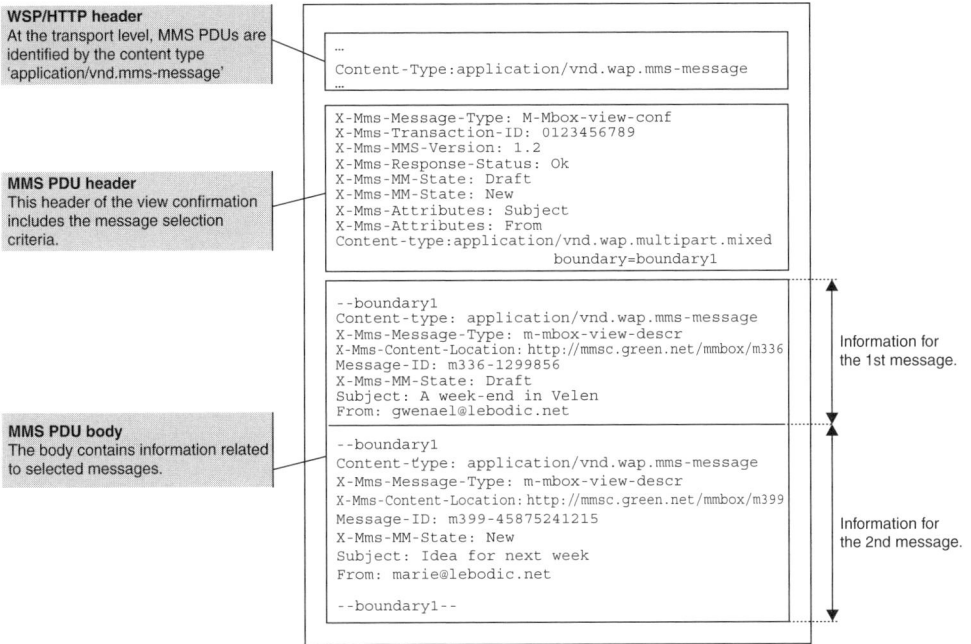

**Figure 6.21**   View confirmation containing information for two selected messages

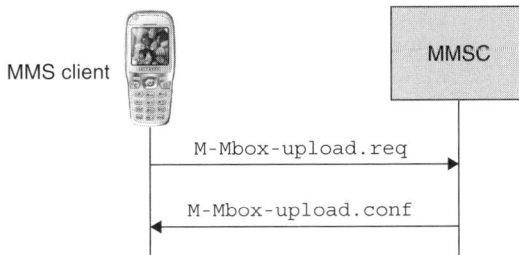

MMS client

MMSC

M-Mbox-upload.req

M-Mbox-upload.conf

**Figure 6.22**   MM1 MMBox upload transaction flow

### 6.2.9 Uploading a Message to the MMBox

From MMS 1.2, the MMS client has the possibility to upload a message in the MMBox (only if the concept of MMBox is supported by the MMSC). The message may be a new message created by the user (e.g. draft message) or a message previously retrieved from the MMSC. The transaction flow for uploading a message in the MMBox is shown in Figure 6.22. The uploading request is named M-Mbox-upload.req and is acknowledged with the M-Mbox-upload.conf confirmation.

Parameters of the M-Mbox-upload.req PDU are described in Table 6.19 and parameters of the M-Mbox-upload.conf PDU are described in Table 6.20. The body of the upload request PDU contains the message to be uploaded in the MMBox.

**Table 6.19**   MM1 MMBox upload request (`M-Mbox-upload.req`)

| Parameter name | Description | From OMA | St. |
|---|---|---|---|
| X-Mms-Message-Type | MMS PDU type. Value: M-Mbox-upload-req | 1.2 | ● |
| X-Mms-Transaction-ID | Unique identifier for the upload transaction. | 1.2 | ● |
| X-Mms-MMS-Version | MMS protocol version such as 1.2. | 1.2 | ● |
| X-Mms-MM-State | MMBox message state. This parameter specifies the state of the uploaded message. If not present, then the default message state is 'draft'. | 1.2 | ○ |
| X-Mms-MM-Flags | MMBox message flags. This parameter indicates the list of flags associated to a message stored in the MMBox. This field may appear multiple times. | 1.2 | ○ |
| Content-Type | Content type of the uploaded message. | 1.2 | ● |

**Table 6.20**   MM1 MMBox upload confirmation (`M-Mbox-upload.conf`)

| Parameter name | Description | From OMA | St. |
|---|---|---|---|
| X-Mms-Message-Type | MMS PDU type. Value: M-Mbox-upload-conf | 1.2 | ● |
| X-Mms-Transaction-ID | Unique identifier for the upload transaction. The same as the one for the corresponding MMBox upload request. | 1.2 | ● |
| X-Mms-MMS-Version | MMS protocol version such as 1.2. | 1.2 | ● |
| X-Mms-Content-Location | This parameter can be present only if the X-Mms-Store-Status is set to 'success'. When available, this parameter indicates a reference of the message in the MMBox. This reference can be used subsequently for retrieving the message or for obtaining information about the message. | 1.2 | ○ |
| X-Mms-Store-Status | MMBox store request status. The status of the upload request such as 'success'. See status codes in Appendix D. | 1.2 | ● |
| X-Mms-Store-Status-Text | MMBox message textual status. Textual description qualifying the value assigned to the X-Mms-Store-Status parameter. | 1.2 | ○ |

In this context, the uploaded message is described with the parameters presented in Table 6.18. The upload confirmation PDU does not contain any body (header only).

It is recommended to use the `M-MBox-View.req` request (instead of the `WSP/HTTP GET.req`) for retrieving a message that has been uploaded to the user's MMBox with the `M-Mbox-upload.req` request.

## 6.2.10 Deleting a Message from the MMBox

From MMS 1.2, the MMS client has the possibility to delete one or more messages from the MMBox with a single transaction (only if the concept of MMBox is supported by the MMSC). The transaction flow for deleting one or more messages from the MMBox is shown in Figure 6.23. The delete request is named `M-Mbox-delete.req` and is acknowledged with the `M-Mbox-delete.conf` confirmation.

Parameters of the `M-Mbox-delete.req` PDU are described in Table 6.21 and parameters of the `M-Mbox-delete.conf` PDU are described in Table 6.22. The delete request and confirmation PDUs do not contain any body (header only).

It may happen that the MMSC is unable to delete several or all messages referenced in the deletion request. In order to warn the requesting MMS client, the MMSC includes in the deletion confirmation, the references (`X-Mms-Content-Location` parameter) of messages that could not be deleted along with relevant error codes (`X-Mms-Response-Status` and `X-Mms-Response-Status-Text` parameters).

## 6.2.11 Parameter Description and Binary Encoding

In the WAP environment, protocol data units are binary encoded for being conveyed over the MM1 interface. Each parameter name has an associated assigned number and each parameter value is encoded according to encoding rules specific to the value

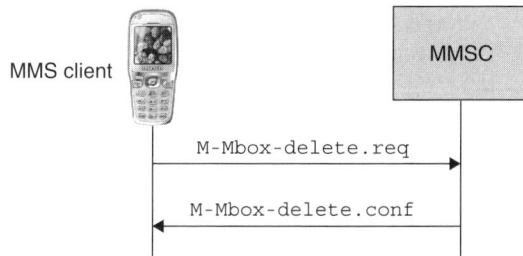

**Figure 6.23** MM1 MMBox delete transaction flow

**Table 6.21** MM1 MMBox delete request (`M-Mbox-delete.req`)

| Parameter name | Description | From OMA | St. |
|---|---|---|---|
| `X-Mms-Message-Type` | MMS PDU type. Value: `M-Mbox-delete-req` | 1.2 | ● |
| `X-Mms-Transaction-ID` | Unique identifier for the delete transaction. | 1.2 | ● |
| `X-Mms-MMS-Version` | MMS protocol version such as 1.2. | 1.2 | ● |
| `X-Mms-Content-Location` | Reference of the message(s) to be deleted. This parameter may appear multiple times if more than one message are to be deleted. | 1.2 | ● |

**Table 6.22**  MM1 MMBox delete confirmation (`M-Mbox-delete.conf`)

| Parameter name | Description | From OMA | St. |
|---|---|---|---|
| `X-Mms-Message-Type` | MMS protocol data unit type. Value: `M-Mbox-delete-conf` | 1.2 | ● |
| `X-Mms-Transaction-ID` | Unique identifier for the delete transaction. The same as the one for the corresponding MMBox delete request. | 1.2 | ● |
| `X-Mms-MMS-Version` | MMS protocol version such as 1.2. | 1.2 | ● |
| `X-Mms-Content-Location` | Reference to the message that could not be deleted by the MMSC. This parameter may appear multiple times if more than one message could not be deleted by the MMSC. Each instance of this parameter is assigned an index number referring to a given status code and an optional error textual description as specified by the two following parameters. | 1.2 | ○ |
| `X-Mms-Response-Status` | MMBox deletion request status. The status of the deletion request. See status codes in Appendix B. Each relevant status code appears only once and is uniquely indexed with a number. | 1.2 | ● |
| `X-Mms-Response-Status-Text` | MMBox message textual status. Textual description qualifying the value assigned to the `X-Mms-Response-Status` parameter. Each relevant status text appears only once and is linked to the corresponding status code with the unique index number. | 1.2 | ○ |

type (e.g. text string, short integer, etc.). In the context of MMS, the following types are used:

- *Enumeration*: A list of predefined values can be assigned to a parameter of type 'enumeration'.
- *Short integer*: A value in the range $0 \ldots 127$ (decimal) can be assigned to a parameter of type 'short integer'. The value is represented with 1 byte for which the most significant bit is set to 1 and the remaining bits represent the integer value.
- *Long integer*: A large integer can be assigned to a parameter of type 'long integer'. The value is formatted according to the following derivation rule:

```
Long-integer-value = Short-length 1*30 Octet
```
where:
`Short-length` is a 1-byte integer representing the number of following bytes (0 to 30 bytes).

`1*30 Octet` is a sequence of up to 30 bytes representing an unsigned integer value with the most significant byte encoded first (big-endian representation).

- *Text string*: Values of type 'text string' contain only US-ASCII characters. The sequence of characters is null terminated.
- *Encoded string*: Values of type 'encoded string' contain US-ASCII characters if the character set is not specified, otherwise these strings are encoded according to UTF-8 [RFC-2279]. The string is formatted according to the following derivation rule:

```
Encoded-string-value = Text-string | Value-length
Char-set Text-string
```

where

`Value-length` is the length of the value in bytes [OMA-WSP]. The value length is either a short length (see long integer derivation rule above) or an integer length spanning over several bytes as defined in [OMA-WSP].
`Char-set` is the character set MIBEnum[2] value registered by Internet Assigned Numbers Authority (IANA).
`Text-string` is a sequence of null-terminated characters.

For instance, the following hexadecimal sequence of bytes represents the UTF-8 encoded string 'Hello':

| | |
|---|---|
| 0x07 | Length of value remaining part (7 bytes – short length) |
| 0xEA | MIBEnum value 0x6A for UTF-8 (106 decimal) |
| 0x48 | Character 'H' |
| 0x65 | Character 'e' |
| 0x6C | Character 'l' |
| 0x6C | Character 'l' |
| 0x6F | Character 'o' |
| 0x00 | Null character – End of string. |

- *Absolute or relative date (GMT)*: A date is either represented in an absolute or relative format according to the following derivation rule:

```
AR-Date-value = Value-length (Absolute-token Date-value |
Relative-token Delta-seconds-value)
```

where

`Value-length` is the length of the value in bytes [OMA-WSP],
`Absolute-token` is the value 0x80,
`Relative-token` is the value 0x81,
`Date-value` is a value of type 'long integer' representing the number of seconds elapsed since 1970-01-01, 00:00:00 GMT,

---

[2] See registered MIBEnum values at http://www.iana.org/assignments/character-sets

`Delta-seconds-value` is a value of type 'long integer' representing the number, of seconds from the reference date and time.

An absolute date is always conveyed as a GMT date. Such a date usually requires to be converted to a local date before it is provided to the user.

- *Complex structure for* `From` *parameter*: Values of this type are assigned to the `From` parameter. They are constructed according to the following derivation rule:

```
From-value = Value-length (Address-present-token Encoded-
string-value | Insert-address-token)
```

where

`Value-length` is the length of the value in bytes,
`Address-present-token` is the value 0x80,
`Insert-address-token` is the value 0x81,
`Encoded-string-value` is a string encoded as shown above.

- *Indexed encoded string*: Values of this type are encoded strings prefixed by either a long integer or a short integer (index) as shown by the following derivation rule:

```
Indexed-encoded-string-value = Value-length Index-value
Encoded-string-value
```

where

`Value-length` is the length of the value in bytes,
`Index-value` is a short or long integer (index) as defined above,
`Encoded-string-value` is an encoded string as defined above.

- *Indexed date*: Values of this type are absolute dates prefixed by either a long or a short integer as shown by the following derivation rule:

```
Indexed-date-value = Value-length Index-value Encoded-
date-value
```

where

`Value-length` is the length of the value in bytes,
`Index-value` is a short or long integer (index) as defined above,
`Encoded-date-value` is a value of type 'long integer' representing the number of seconds elapsed since 1970-01-01, 00:00:00 GMT.

- *MMBox size*: This type is mainly used for representing MMBox totals and quotas. Values of this type represent a size expressed in terms of number of messages or number of bytes as shown by the following derivation rule:

```
Mbox-size-value = Value-length (Message-token | Size-token)
Integer-Value
```

where

`Value-length` is the length of the value in bytes,
`Message-token` is the value 0x80 (size is expressed in number of messages),
`Size-token` is the value 0x81 (size is expressed in number of bytes),
`Integer-Value` is a short or long integer representing the size (in terms of messages or bytes).

- *Flag command*: A flag command instructs the MMSC to add, remove or filter a flag for a message stored in the MMBox. A flag command is structured as shown in the following derivation rule:

```
Flag-command-value = Value-length (Add-token | Remove-
token | Filter-token) Encoded-string-value
```

where

`Value-length` is the length of the value in bytes,
`Add-token` is the value 0x80 (command is 'add a flag'),
`Remove-token` is the value 0x81 (command is 'remove flag'),
`Filter-token` is the value 0x82 (command is 'filter flag'),
`Encoded-string-value` is a US-ASCII or UTF-8 string representing the flag.

- *Element descriptor*: A value of type element descriptor describes the main structure of a message (multipart mixed, related, etc.), available from MMS 1.2. Such a value is structured as shown with the following derivation rules:

```
Element-Descriptor-value = Value-length Content-Reference-
value *(Parameter) Parameter = Parameter-name Parameter-value
```

where

`Value-length` is the length of the value in bytes,
`Content-Reference-Value` is a text string representing the element reference,
`Parameter-name` is the name of the parameter as a text string (US-ASCII) or short integer (assigned number – see parameter binary encoding in Table 6.23),
`Parameter-value` is the value assigned to the corresponding parameter (see possible values for each parameter in Table 6.23).

In the encoding of the PDU header parameters, the order of the parameters is not significant, except that `X-Mms-Message-Type`, `X-Mms-Transaction-ID` (when present) and `X-Mms-MMS-Version` parameters must be at the beginning of the PDU header, in that order, and if the PDU contains a multimedia message in the PDU body, then the `Content-Type` parameter must be the last header parameter, followed by message contents.

The list of parameters that can be included in PDUs invoked over the MM1 interface is provided in Table 6.23.

**Table 6.23**   MM1 PDU parameters with binary encoding

| Parameter name | A.N.[a] | B.E.[b] | Values | Binary encoding |
|---|---|---|---|---|
| Bcc | 0x01 | 0x81 | US-ASCII or UTF-8 string | Type: encoded string |
| Cc | 0x02 | 0x82 | US-ASCII or UTF-8 string | Type: encoded string |
| Content-Type | 0x04 | 0x84 | Multipart mixed/related | Type: multipart as defined in [OMA-WSP] |
| Date | 0x05 | 0x85 | In seconds since 1970-01-01, 00:00:00 GMT | Type: long integer |
| From | 0x09 | 0x89 | US-ASCII or UTF-8 string or 'insert token'. | Type: complex structure for From parameter. |
| Message-ID | 0x0B | 0x8B | US-ASCII string | Type: text string |
| Subject | 0x16 | 0x96 | US-ASCII or UTF-8 string | Type: encoded string |
| To | 0x17 | 0x97 | US-ASCII or UTF-8 string | Type: encoded string |
| X-Mms-Attributes | 0x28 | 0xA8 | | Type: enumeration |
| | | | Bcc | 0x81 |
| | | | Cc | 0x82 |
| | | | Content | 0xAE |
| | | | Content-Type | 0x84 |
| | | | Date | 0x85 |
| | | | X-Mms-Delivery-Report | 0x86 |
| | | | X-Mms-Delivery-Time | 0x87 |
| | | | X-Mms-Expiry | 0x88 |
| | | | From | 0x09 |
| | | | X-Mms-Message-Class | 0x8A |
| | | | Message-ID | 0x8B |
| | | | X-Mms-Message-Size | 0x8E |
| | | | X-Mms-Priority | 0x8F |
| | | | X-Mms-Read-Report | 0x90 |
| | | | Subject | 0x96 |
| | | | To | 0x97 |
| | | | X-Mms-Reply-Charging | 0x9C |
| | | | X-Mms-Reply-Charging-ID | 0x9E |
| | | | X-Mms-Reply-Charging-Deadline | 0x9D |
| | | | X-Mms-Reply-Charging-Size | 0x9F |
| | | | X-Mms-Previously-Sent-By | 0xA0 |
| | | | X-Mms-Previously-Sent-Date | 0xA1 |

**Table 6.23**   (*continued*)

| Parameter name | A.N.[a] | B.E.[b] | Values | Binary encoding |
|---|---|---|---|---|
| | | | Additional-headers | 0xB0 |
| X-Mms-Content-Location | 0x03 | 0x83 | US-ASCII string | Type: text string[c] |
| X-Mms-Delivery-Report | 0x06 | 0x86 | | Type: enumeration |
| | | | Yes | 0x80 |
| | | | No | 0x81 |
| X-Mms-Delivery-Time | 0x07 | 0x87 | Absolute time or relative time from submission time. | Type: absolute or relative date |
| X-Mms-Distribution-Indicator | 0x31 | 0xB1 | | Type: enumeration |
| | | | Yes | 0x80 |
| | | | No | 0x81 |
| X-Mms-Element-Descriptor | 0x32 | 0xB2 | | Type: element descriptor |
| X-Mms-Expiry | 0x08 | 0x88 | Absolute time or relative time from submission time. | Type: absolute or relative date |
| X-Mms-Mbox-Quotas | 0x2C | 0xAC | Quotas of the MMBox. | Type: MMBox size |
| X-Mms-Mbox-Totals | 0x2A | 0xAA | Number of messages or bytes stored in the MMBox. | Type: MMBox size |
| X-Mms-Message-Class | 0x0A | 0x8A | Token text or | Type: enumeration |
| | | | Personal | 0x80 |
| | | | Advertisement | 0x81 |
| | | | Informational | 0x82 |
| | | | Auto | 0x83 |
| X-Mms-Message-Count | 0x2D | 0xAD | Number of messages | Type: short or long integer |
| X-Mms-Message-Size | 0x0E | 0x8E | Expressed in bytes | Type: long integer |
| X-Mms-Message-Type | 0x0C | 0x8C | | Type: enumeration |
| | | | M-send-req | 0x80 |
| | | | M-send-conf | 0x81 |
| | | | M-notification-ind | 0x82 |
| | | | M-notifyresp-ind | 0x83 |
| | | | M-retrieve-conf | 0x84 |
| | | | M-acknowledge-ind | 0x85 |

(*continued overleaf*)

**Table 6.23** (*continued*)

| Parameter name | A.N.[a] | B.E.[b] | Values | Binary encoding |
|---|---|---|---|---|
| | | | M-delivery-ind | 0x86 |
| | | | M-read-rec-ind | 0x87 |
| | | | M-read-orig-ind | 0x88 |
| | | | M-forward-req | 0x89 |
| | | | M-forward-conf | 0x8A |
| | | | M-mbox-store-req | 0x8B |
| | | | M-mbox-store-conf | 0x8C |
| | | | M-mbox-view-req | 0x8D |
| | | | M-mbox-view-conf | 0x8E |
| | | | M-mbox-upload-req | 0x8F |
| | | | M-mbox-upload-conf | 0x90 |
| | | | M-mbox-delete-req | 0x91 |
| | | | M-mbox-delete-conf | 0x92 |
| | | | M-mbox-descr | 0x93 |
| X-Mms-MM-Flags | 0x24 | 0xA4 | Add, remove or filter flag command | Type: flag command |
| X-Mms-MM-State | 0x23 | 0xA3 | | Type: enumeration |
| | | | Draft | 0x80 |
| | | | Sent | 0x81 |
| | | | New | 0x82 |
| | | | Retrieved | 0x83 |
| | | | Forwarded | 0x84 |
| X-Mms-MMS-Version | 0x0D | 0x8D | | Type: short integer |
| | | | MMS 1.0 | 0x90 |
| | | | MMS 1.1 | 0x91 |
| | | | MMS 1.2 | 0x92 |
| X-Mms-Previously-Sent-By | 0x20 | 0xA0 | A US-ASCII or UTF-8 string prefixed with a 'forward count'. | Type: indexed encoded string |
| X-Mms-Previously-Sent-Date | 0x21 | 0xA1 | An absolute date prefixed with a 'forward count'. | Type: indexed date |
| X-Mms-Priority | 0x0F | 0x8F | | Type: enumeration |
| | | | Low | 0x80 |
| | | | Medium | 0x81 |
| | | | High | 0x82 |
| X-Mms-Quotas | 0x2B | 0xAB | | Type: enumeration |
| | | | Yes | 0x80 |
| | | | No | 0x81 |
| X-Mms-Read-Report | 0x10 | 0x90 | | Type: enumeration |
| | | | Yes | 0x80 |
| | | | No | 0x81 |

**Table 6.23**   (*continued*)

| Parameter name | A.N.[a] | B.E.[b] | Values | Binary encoding |
|---|---|---|---|---|
| X-Mms-Read-Status | 0x1B | 0x9B | | Type: enumeration |
| | | | Read | 0x80 |
| | | | Deleted without being read | 0x81 |
| X-Mms-Reply-Charging | 0x1C | 0x9C | | Type: enumeration |
| | | | Requested | 0x80 |
| | | | Requested text only | 0x81 |
| | | | Accepted | 0x82 |
| | | | Accepted text only | 0x83 |
| X-Mms-Reply-Charging-Deadline | 0x1D | 0x9D | Absolute time or relative time from submission time. | Type: absolute or related date |
| X-Mms-Reply-Charging-ID | 0x1E | 0x9E | US-ASCII string | Type: text string |
| X-Mms-Reply-Charging-Size | 0x1F | 0x9F | Number of bytes. | Type: long integer |
| X-Mms-Report-Allowed | 0x11 | 0x91 | | Type: enumeration |
| | | | Yes | 0x81 |
| | | | No | 0x82 |
| X-Mms-Response-Status | 0x12 | 0x92 | See error codes in Appendix B. | Type: enumeration[d] |
| X-Mms-Response-Text | 0x13 | 0x93 | US-ASCII or UTF-8 string | Type: encoded string[d] |
| X-Mms-Retrieve-Status | 0x19 | 0x99 | See error codes in Appendix C. | Type: enumeration |
| X-Mms-Retrieve-Text | 0x1A | 0x9A | US-ASCII or UTF-8 string | Type: encoded string |
| X-Mms-Sender-Visibility | 0x14 | 0x94 | | Type: enumeration |
| | | | Hide | 0x80 |
| | | | Show | 0x81 |
| X-Mms-Status | 0x15 | 0x95 | | Type: enumeration |
| | | | Expired | 0x80 |
| | | | Retrieved | 0x81 |
| | | | Rejected | 0x82 |
| | | | Deferred | 0x83 |
| | | | Unrecognized | 0x84 |
| | | | Indeterminate | 0x85 |
| | | | Forwarded | 0x86 |
| X-Mms-Store | 0x22 | 0xA2 | | Type: enumeration |

(*continued overleaf*)

**Table 6.23**  (*continued*)

| Parameter name | A.N.[a] | B.E.[b] | Values | Binary encoding |
|---|---|---|---|---|
| | | | Yes | 0x80 |
| | | | No | 0x81 |
| X-Mms-Stored | 0x27 | 0xA7 | | Type: enumeration |
| | | | Yes | 0x80 |
| | | | No | 0x81 |
| X-Mms-Store-Status | 0x25 | 0xA5 | See error codes in Appendix D. | Type: enumeration |
| X-Mms-Store-Status-Text | 0x26 | 0xA6 | US-ASCII string | Type: text string |
| X-Mms-Totals | 0x29 | 0xA9 | | Type: enumeration |
| | | | Yes | 0x80 |
| | | | No | 0x81 |
| X-Mms-Transaction-ID | 0x18 | 0x98 | US-ASCII string | Type: text string |

[a] A.N. stands for Assigned Number.

[b] B.E. stands for Binary Encoding.

[c] In the M-Mbox-dalete.conf PDU, parameter X-Mms-Content-Location is encoded according to the derivation rule: Value-length Status-count-value Encoded string-value where Status-count-value is a unique index number referring to the relevant status code (X-Mms-Response-Status).

[d] In the M-Mbox-delete.conf PDU, X-Mms-Response-Status and X-Mms-Response-Text parameters are indexed to provide a list of unique relevant status codes and corresponding text descriptions.

**Table 6.24**  Maximum lengths for MM1 parameters

| Parameter name | Maximum length |
|---|---|
| Content-ID | 100 characters |
| Content-Location | 100 characters |
| Message-ID | 40 characters |
| Subject | 40 characters[a] |
| X-Mms-Content-Location | 100 characters |
| X-Mms-Response-Text | 30 characters |
| X-Mms-Transaction-ID | 40 characters |

[a] For the subject parameter, 40 characters represent either 40 US-ASCII characters or 40 complex glyphs (e.g. Chinese pictograms).

To guarantee interoperability, OMA recommends maximum lengths for several PDU parameters of the MM1 interface as shown in Table 6.24 [OMA-MMS-Conf].

## 6.3 MM2 Interface, Internal MMSC Interface

The MM2 interface is an MMSC internal interface. The interface is required if MMS Relay and MMS Server are provided as two separate entities (e.g. developed by two different manufacturers). However, available commercial implementations usually combine MMS Relay and MMS Server into a single entity, called MMSC. In this situation, the MM2 interface is implemented in a proprietary manner. At the time of writing, no technical realization of the MM2 interface had been specified by standardization organizations.

## 6.4 MM3 Interface, MMSC–External Servers

The MM3 interface allows the MMSC to exchange messages with external servers such as email servers and SMS centres (SMSCs). This interface is typically based on existing IP-based transport protocols (e.g. HTTP or SMTP).

When sending a message to an external messaging system, the MMSC converts the multimedia message into an appropriate format supported by the external messaging system. For instance, the exchange of a message between an MMSC and an Internet email server can be performed by converting the multimedia message from its MM1 binary multipart representation into a text-based MIME representation for transfer over SMTP. In order to receive a message from an external messaging system, the MMSC converts incoming messages to a format supported by receiving MMS clients.

Several mechanisms can be used for discovering incoming messages from external messaging servers. These mechanisms include the following:

- Forwarding of the message from the external messaging server to the MMSC.
- The external messaging server notifies the MMSC that a message is waiting for retrieval. In this configuration, it is the responsibility of either the MMSC or the recipient MMS client to explicitly retrieve the message.
- The MMSC can periodically poll the external messaging server for incoming messages.

Mobile users are often able to send multimedia messages to Internet users. Owing to obvious billing issues, it is usually not possible for Internet users to send messages (received as multimedia messages) to mobile users if the applicable billing model is the one where the sender pays for message delivery.

## 6.5 MM4 Interface, MMSC–MMSC

The MM4 interface allows interactions between two MMSCs. Such interactions are required in the situation in which multimedia messages and associated reports are exchanged between users attached to distinct MMSEs. In this context, a multimedia message is always delivered to the recipient via the recipient MMSC, distinct from the originator MMSC. Note that this mechanism for exchanging messages between MMS users differs from the one usually in place for SMS. For the exchange of an

**Table 6.25** List of MM4 transactions/PDUs

| Transaction | PDU name | Description | From 3GPP |
|---|---|---|---|
| Routing forward a message. | `MM4_forward.REQ` | Message forward request | Rel-4 |
| | `MM4_forward.RES` | Message forward confirmation | |
| Routing forward a delivery report | `MM4_delivery_report.REQ` | Request for delivery report forward | Rel-4 |
| | `MM4_delivery_report.RES` | Confirmation for delivery report forward | |
| Routing forward a read report | `MM4_read_reply_report.REQ` | Request for read report forward | Rel-4 |
| | `MM4_read_reply_report.RES` | Confirmation for read-report forward | |

SMS message, the originator SMSC is in charge of delivering directly the message to the recipient(s)[3].

Table 6.25 lists the three transactions that can occur over the MM4 interface (6 PDUs).

At the time of writing, the only standardized technical realization available for the MM4 interface was the one defined by 3GPP in [3GPP-23.140] (from Release 4). In this technical realization, transactions specified for the MM4 interface are conveyed over the Simple Mail Transfer Protocol (SMTP) as shown in Figure 6.24.

As for the MM1 interface, each transaction in the MM4 interface is usually composed of a request and a response/confirmation. The 3GPP naming convention for naming requests and responses is used in this book for the MM4 interface (see Section 6.1).

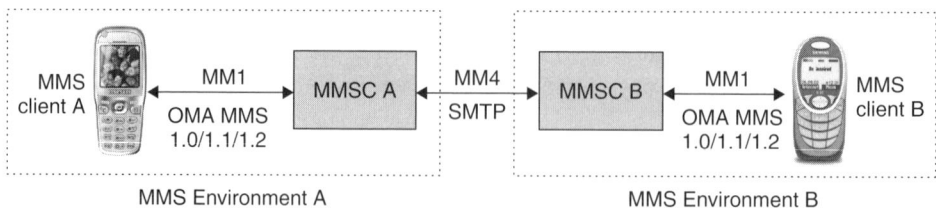

**Figure 6.24** Interworking between two distinct MMSEs

---

[3] In rare cases, it happens that the delivery of an SMS message is performed by the recipient SMSC. This configuration is implemented when two networks cannot interwork directly because they rely on incompatible transport technologies.

### 6.5.1 Introduction to SMTP

SMTP is a basic protocol for exchanging messages between Mail User Agents (MUAs). MUA is the name given to applications in charge of managing email messages exchanged over the Internet. In the Internet domain, SMTP has become the de facto email transfer protocol for exchanging messages. MUAs have similar responsibilities as those of MMS clients in an MMSE. Specifications of SMTP have been published by the Internet Engineering Task Force (IETF) in [RFC-2821].

The SMTP model is based on an interconnection of Message Transfer Agents (MTAs). In a typical SMTP transaction, an MTA is the sender-SMTP if it originates the SMTP commands, or the receiver-SMTP if it handles the received SMTP commands. An MTA can usually play both roles: the sender-SMTP client and the receiver-SMTP server. SMTP defines how senders and receivers can initiate a transfer session, can transfer messages(s) over a session and can tear down an open session. Note that how the message is physically transferred from the sender to the receiver is not defined as part of the SMTP specifications (see Box 6.2 below). SMTP only defines the set of commands, and corresponding responses, for controlling the transfer of messages over sessions.

---

**BOX 6.2 Interworking between distinct environments, GSMA recommendations**

Ability to exchange messages between distinct environments is crucial for the success of MMS. To enable this, 3GPP has identified SMTP as a suitable transport protocol for the realization of the interface bridging MMS centres belonging to different MMS providers.

Furthermore, the GSM Association (GSMA) has published recommendations to ensure an efficient interworking over the MM4 interface [GSMA-IR.52]. The interconnection between two mobile networks for the realization of the MM4 interface can be performed over the public Internet (with secure connections) or over direct leased lines such as Frame relay or Asynchronous Transfer Mode (ATM). However, GSMA recommends the use of an alternative solution based on GRX [GSMA-IR.34] as an interconnection and transmission network for MMS traffic. GRX enables multimedia messages to be routed as IP-based traffic between two distinct mobile networks.

In the roaming scenario, the roaming user still uses the home MMSC in order to send and retrieve multimedia messages. This means that a roaming user is not required to change the MMS device settings in order to have access to the service. The visited network is not required to provide support for MMS, but the user should be able to establish a data connection (e.g. GSM Circuit Switched Data (CSD) or GPRS) in order to have access to the service.

---

SMTP is a stateful protocol, meaning that the sender and the receiver involved in operations over a session maintain a current context for a session. Consequently, commands requested over SMTP have different results according to the session state.

In the context of MMS, SMTP is used to transfer multimedia messages, delivery reports and read reports between MMSCs. For this purpose, originator and recipient MMSCs are MTAs playing the roles of sender-SMTP and receiver-SMTP respectively.

An SMTP command is a four-letter command such as HELO or DATA. The response to such a command consists of a three-digit code followed by some optional human readable text. The status code is formatted according to the convention shown in Figure 6.25.

The minimum set of SMTP commands that should be supported by MMSCs is the following:

- HELO: This command (abbreviation for 'Hello') is used for initiating a session.
- QUIT: This command is used to tear down a session.
- MAIL: This command tells the SMTP-receiver that a message transfer is starting and that all state tables and buffers should be re-initialized. This command has one parameter named FROM, which identifies the address of the message originator.
- RCPT: This command (abbreviation for 'Recipient') has one parameter, named TO, which identifies the address for one of the message recipients. If the message is sent to several recipients, then this command may be executed multiple times for one single message transfer.
- DATA: This command is used for transferring the message itself.

Phone numbers and email addresses can be used for addressing recipients in an MMSE. With SMTP, only email addresses are supported for routing purpose. Consequently, phone numbers specified as part of MAIL and RCPT parameters, need to be adapted from original and recipient MMS client addresses. For instance, the phone number +33672873762 for an originator MMS client is converted to a Fully Qualified Domain Name (FQDN) email address as illustrated in Figure 6.26.

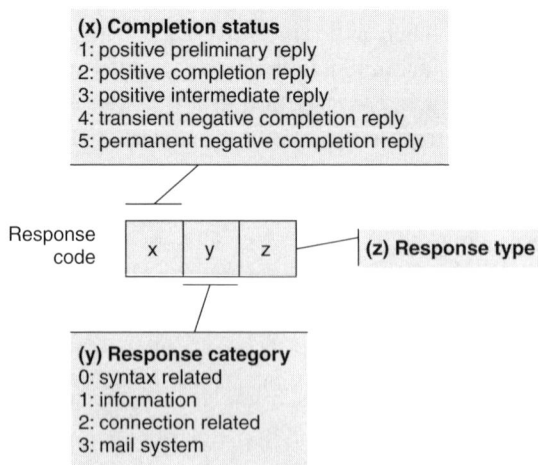

**Figure 6.25**  Structure of an SMTP response

**MMS address**
The MMS address is composed of the original MMS address along with its type (in this example PLMN).

**Domain name**
The domain name identifies the MMS environment (usually the operator domain).

+33667287376/TYPE=PLMN@mmse.operator.net

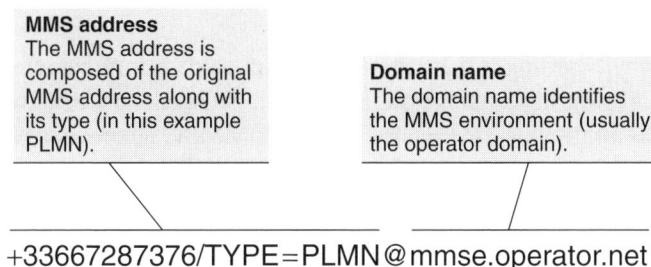

**Figure 6.26** SMTP address conversion

For sending a message to an external MMSE, the originator MMSC needs to resolve the recipient MMSC domain name to an IP address. If the recipient address is an email address, then the originator MMSC can obtain the recipient MMSC IP address by interrogating a Domain Name Server (DNS) with the recipient address. If the recipient address is in the form of a phone number, then two resolution methods have been identified by 3GPP in [3GPP-23.140] Release 4:

- *DNS-ENUM-based method*: IETF has identified a method for locating a device associated with a phone number with a DNS [RFC-2916]. This method, known as DNS-ENUM, allows DNS records (including the associated email address) to be retrieved using the recipient phone number as a record key. With the email address, the originator MMSC determines the recipient MMSC IP address with the normal DNS method as described above.
- *IMSI-based method*: This method for identifying the recipient MMSC IP address complies with the Mobile Number Portability (MNP) requirements. The MMSC interrogates the recipient Home Location Register (HLR) in order to get the International Mobile Subscriber Identity (IMSI) associated with the recipient phone number (via the MM5 interface). From the IMSI, the originator MMSC extracts the Mobile Network Code (MNC) and the Mobile Country Code (MCC). With the MNC and MCC, the originator MMSC obtains the recipient MMSC FQDN by interrogating a local database (look-up table). With the MMSC FQDN, the originator MMSC retrieves the MMSC IP address by interrogating a DNS.

MMSCs can support the following SMTP extensions:

- SMTP service extension for message size declaration.
- SMTP service extension for 8-bit MIME transport.

The support of additional commands is also possible, but their use has not yet been covered in 3GPP technical specifications. An example for transferring a message over SMTP between two distinct MMSEs is provided in Section 6.5.5. The following sections present transaction flows for message forward, delivery-report forward and read-report forward.

### 6.5.2 Routing Forward a Message

The request for forwarding a message (`MM4_forward.REQ` request) over the MM4 interface allows a multimedia message to be transferred between two MMSEs. If requested by the originator MMSC, the recipient MMSC acknowledges the message forward request with a message forward response (`MM4_forward.RES` response). The transaction flow for the transfer of a message over the MM4 interface is shown in Figure 6.27.

The `MM4_forward.REQ` request is composed of parameters listed in Table 6.26. If requested by the originator MMSC, the recipient MMSC acknowledges the message forward request with a message forward response (`MM4_forward.RES` response). The response is composed of parameters listed in Table 6.27.

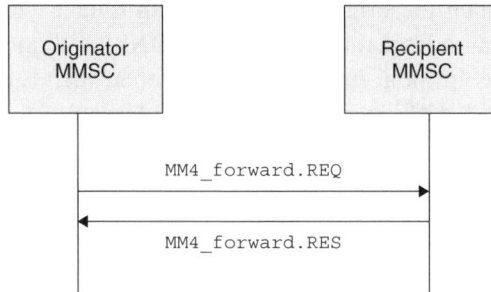

**Figure 6.27**    MM4 message forward

**Table 6.26**    MM4 message forward request (`MM4_forward.REQ`)

| Parameter name | Description | From 3GPP | St. |
|---|---|---|---|
| X-Mms-3GPP-MMS-Version | 3GPP MMS Version, MMS Version of the MMSC – Type: string – Example: `5.2.0` | Rel-4 | ● |
| X-Mms-Message-Type | Type of MM4 operation (request for the forward of a message) – Type: string – Value: `MM4_forward.REQ` | Rel-4 | ● |
| X-Mms-Transaction-ID | Identifier of the forward transaction – Type: string | Rel-4 | ● |
| X-Mms-Message-ID | Identifier of the multimedia message being forwarded – Type: string | Rel-4 | ● |
| To, Cc, Bcc | Recipient address(es), address of the recipient(s) of the original message – Type: string | Rel-4 | ● |
| From | Sender address, address of the sender of the original message – Type: string | Rel-4 | ● |
| Subject | Message subject – Type: string – Condition (A) | Rel-4 | © |
| X-Mms-Message-Class | Message class – Type: string – Condition (A) – Values: `Personal`, `Advertisement`, `Information` or `Auto` | Rel-4 | © |

**Table 6.26** (*continued*)

| Parameter name | Description | From 3GPP | St. |
|---|---|---|---|
| Date | Date and time when the original message was submitted – Type: date | Rel-4 | ● |
| X-Mms-Expiry | Time of expiry of the message – Type: date or duration – Condition (A) | Rel-4 | © |
| X-Mms-Delivery-Report | Whether or not a delivery report is requested – Condition (A) – Values: Yes or No | Rel-4 | © |
| X-Mms-Read-Reply | Whether or not a read report is requested – Condition (A) – Values: Yes or No | Rel-4 | © |
| X-Mms-Priority | Priority of the message being forwarded – Condition (A) – Values: Low, Normal or High | Rel-4 | © |
| X-Mms-Sender-Visibility | Whether or not the sender requested sender details to be hidden from recipients – Condition (A) – Values: Hide or Show | Rel-4 | © |
| X-Mms-Forward-Counter | Counter indicating how many times the message has been forwarded – Condition (A) – Type: Integer | Rel-4 | © |
| X-Mms-Previously-Sent-By | Address(es) of MMS clients that have handled (submitted or forwarded) the message prior to the manipulation by the MMS client whose address is assigned to the From parameter – Type: string with index – Example: 1, armel@armorepro.com 2, gwenael@lebodic.net | Rel-4 | O |
| X-Mms-Previously-Sent-Date-and-Time | Date and time when the message was handled Type: date with index – Example: 1, Mon Jan 21 09:45:33 2003 2, Wed Jan 23 18:06:21 2003 | Rel-4 | O |
| X-Mms-Ack-Request | Whether or not an acknowledgement for the forward request is requested – Values: Yes or No | Rel-4 | O |
| Sender | Originator address as determined by the SMTP MAIL FROM command – Type: string | Rel-4 | ● |
| X-Mms-Originator-System | System address to which the requested forward response should be sent – Condition (B) – Type: string | Rel-4 | © |
| Message-ID | Each SMTP request/response has a unique reference assigned to the Message-ID parameter – Type: string | Rel-4 | ● |
| Content-Type | Message content type – Type: string | Rel-4 | ● |
| Message body | Message content – Condition (A) | Rel-4 | © |

Conditions: (A) Available only if provided by the originator MMS client.
(B) Required if a forward response is requested.

**Table 6.27**  MM4 message forward response (`MM4_forward.RES`)

| Parameter name | Description | From 3GPP | St. |
|---|---|---|---|
| `X-Mms-3GPP-MMS-Version` | 3GPP MMS Version, MMS Version of the MMSC – Type: string – Example: `5.2.0` | Rel-4 | ● |
| `X-Mms-Message-Type` | Type of MM4 operation (response for the forward of a message) – Type: string – Value: `MM4_forward.RES` | Rel-4 | ● |
| `X-Mms-Transaction-ID` | Identifier of the forward transaction – Type: string | Rel-4 | ● |
| `X-Mms-Message-ID` | Identifier of the multimedia message being forwarded – Type: string | Rel-4 | ● |
| `X-Mms-Request-Status-Code` | Status code of the request to forward the message – Values: `Ok` or error codes defined in Appendix E. | Rel-4 | ● |
| `X-Mms-Status-Text` | Optional status text – Type: string | Rel-4 | ○ |
| `Sender` | System address, address of the recipient MMSC – Type: string | Rel-4 | ● |
| `To` | System address, address of the originator MMSC – Type: string | Rel-4 | ● |
| `Message-ID` | Each SMTP request/response has a unique reference assigned to the `Message-ID` parameter – Type: string | Rel-4 | ● |
| `Date` | Date provided by the recipient MMSC – Type: string | Rel-4 | ● |

Unlike forward requests, the addressing of message forward responses is related to neither the message originator nor the message recipient. Instead, the addressing of a forward response is related to special system addresses. The value to be assigned to the `To` parameter of the response is the value assigned to the `X-MMS-Originator-System` parameter of the corresponding forward request (usually a special system address identifying the originator MMSC). The value to be assigned to the `Sender` parameter is a special address identifying the recipient MMSC. It is suggested that special system addresses should be formatted in the form

```
system-user@mms-relay-host.operatorX.net
```

If the forward request has been processed without error by the recipient MMSC, the following value is assigned to the request status code parameter (`X-Mms-Request-Status-Code` parameter).

- `Ok`: This status code indicates that the corresponding request has been processed without errors.

If errors occurred during the processing of the forward request, the codes listed in Appendix E can be assigned to the request status code parameter of the corresponding response.

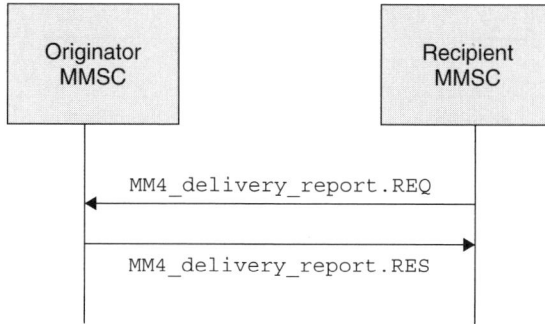

**Figure 6.28** MM4 delivery report

### 6.5.3 Routing Forward a Delivery Report

The request for forwarding a delivery report (`MM4_delivery_report.REQ` request) over the MM4 interface allows the transfer of a delivery report between two MMSEs. If requested by the recipient MMSC, the originator MMSC acknowledges the request with a response (`MM4_delivery_report.RES` response). The transaction flow for the transfer of a delivery report over the MM4 interface is shown in Figure 6.28.

The different delivery states that can be reported over the MM4 interface are as follows:

- Expired
- Retrieved
- Deferred
- Indeterminate
- Forwarded
- Unrecognized.

The `MM4_delivery_report.REQ` request is composed of parameters listed in Table 6.28. The `MM4_delivery_report.RES` response is composed of parameters listed in Table 6.29.

For the response, the value to be assigned to the `To` parameter is the value assigned to the `Sender` parameter of the corresponding forward request. The value to be assigned to the `Sender` parameter is the system address of the recipient MMSC.

### 6.5.4 Routing Forward a Read Report

The request for forwarding a read report (`MM4_read_reply_report.REQ` request) over the MM4 interface allows a read report to be transferred between two distinct MMSEs. If requested by the recipient MMSC, the originator MMSC acknowledges the request with a response (`MM4_read_reply_report.RES` response). The transaction flow for the transfer of a read report over the MM4 interface is shown in Figure 6.29.

**Table 6.28**  MM4 delivery report forward request (`MM4_delivery_report.REQ`)

| Parameter name | Description | From 3GPP | St. |
|---|---|---|---|
| X-Mms-3GPP-MMS-Version | 3GPP MMS – Version, MMS Version of the MMSC – Type: string – Example: `5.2.0` | Rel-4 | ● |
| X-Mms-Message-Type | Type of MM4 operation (request for the forward of a delivery report) – Type: string – Value: `MM4_delivery_report.REQ` | Rel-4 | ● |
| X-Mms-Transaction-ID | Identifier of the forward transaction – Type: string | Rel-4 | ● |
| X-Mms-Message-ID | Identifier of the corresponding multimedia message – Type: string | Rel-4 | ● |
| From | Recipient address, address of the recipient of the original message – Type: string | Rel-4 | ● |
| To | Sender address, address of the sender of the original message – Type: string | Rel-4 | ● |
| Date | Message date and time, Date and time when the message was handled (retrieved, expired, rejected, etc.) – Type: date | Rel-4 | ● |
| X-Mms-Ack-Request | Request for an acknowledgement – Whether or not an acknowledgement of the forward request is requested – Values: `Yes` or `No` | Rel-4 | ○ |
| X-Mms-MM-Status-Code | Message status code, Status of the corresponding multimedia message – Type: string – Values: `Expired`, `Retrieved`, `Deferred`, `Indeterminate`, `Forwarded` or `Unrecognised`. | Rel-4 | ● |
| X-Mms-Status-Text | Text corresponding to the status code – Type: string | Rel-4 | ○ |
| Sender | System address, address to which the requested response should be sent – Type: string | Rel-4 | ● |
| Message-ID | Each SMTP request/response has a unique reference assigned to the `Message-ID` parameter – Type: string | Rel-4 | ● |

The different read states that can be reported over the MM4 interface are as follows:

- Read
- Deleted without being read.

The `MM4_read_reply_report.REQ` request is composed of parameters listed in Table 6.30. The `MM4_read_reply_report.RES` response is composed of parameters listed in Table 6.31.

**Table 6.29** MM4 delivery report forward response (`MM4_delivery_report.RES`)

| Parameter name | Description | From 3GPP | St. |
|---|---|---|---|
| `X-Mms-3GPP-MMS-Version` | 3GPP MMS – Version, MMS Version of the MMSC – Type: string – Example: `5.2.0` | Rel-4 | ● |
| `X-Mms-Message-Type` | Type of MM4 operation (response for the forward of a delivery report) – Type: string – Value: `MM4_delivery_report.RES` | Rel-4 | ● |
| `X-Mms-Transaction-ID` | Identifier of the forward transaction – Type: string | Rel-4 | ● |
| `X-Mms-Message-ID` | Identifier of the corresponding multimedia message – Type: string | Rel-4 | ● |
| `X-Mms-Request-Status-Code` | Status code of the request to forward the delivery report – Values: `Ok` or error codes defined in Appendix E. | Rel-4 | ● |
| `X-Mms-Status-Text` | Optional status text – Type: string | Rel-4 | ○ |
| `Sender` | System address, address of the recipient MMSC – Type: string | Rel-4 | ● |
| `To` | System address, address of the originator MMSC – Type: string | Rel-4 | ● |
| `Message-ID` | Each SMTP request/response has a unique reference assigned to the `Message-ID` parameter – Type: string | Rel-4 | ● |
| `Date` | Date provided by the recipient MMSC – Type: string | Rel-4 | ● |

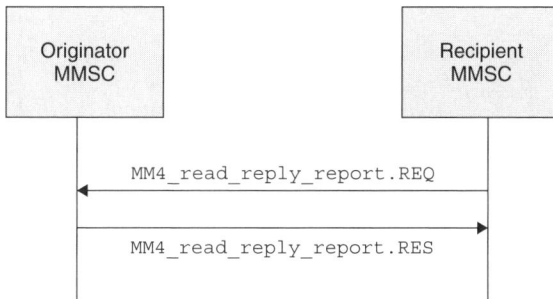

**Figure 6.29** MM4 read report

For the response, the value to be assigned to the `To` parameter is the value assigned to the `Sender` parameter of the corresponding forward request. The value to be assigned to the `Sender` parameter is the system address of the recipient MMSC.

### 6.5.5 Example for Message Transfer with SMTP

Figure 6.30 shows the sequence of SMTP instructions required for (1) opening an SMTP session, (2) transferring a message and (3) tearing down the session. Note that

```
Receiver>  220 MMSE.OPERATORB.NET SMTP Ready
Sender>    HELO MMSE.OPERATORA.NET
Receiver>  250 MMSE.OPERATORB.NET

Sender>    MAIL FROM:<+3367287376/TYPE=PLMN@MMSE.OPERATORA.NET>
Receiver>  250 Ok
Sender>    RCPT TO:<+3364728737 4/TYPE=PLMN@MMSE.OPERATORB.NET>
Receiver>  250 Ok
Sender>    DATA
Receiver>  354 Start message input; end with <CLRF>.<CLRF>
Sender>    X-MMS-3GPP-MMS-version: 4.2.0
Sender>    X-MMS-Message-Type: MM4_forward.REQ
Sender>    X-MMS-Transaction-ID:"634"
Sender>    X-MMS-Message-ID:"566"
Sender>    Date: Wed, 16 May 2001 10:56:00 +0800
Sender>    From: +3367287376/TYPE=PLMN
Sender>    To: +3364728737 4/TYPE=PLMN
Sender>    Subject: A message with text only
Sender>    Content-type: text/plain
Sender>    This is the message content
Sender>    .
Receiver>  250 Ok

Sender>    QUIT
Receiver>  221 MMSE.OPERATORB.NET Service closing transmission session
```

**① Session initiation**
This sequence of instructions initiates a session for transferring one or more messages or reports.

**② Message transfer**
This sequence of instructions enables the message transfer from the sender to the receiver.

**③ Session tear-down**
This sequence of instructions indicates to the receiver that the session is to be terminated.

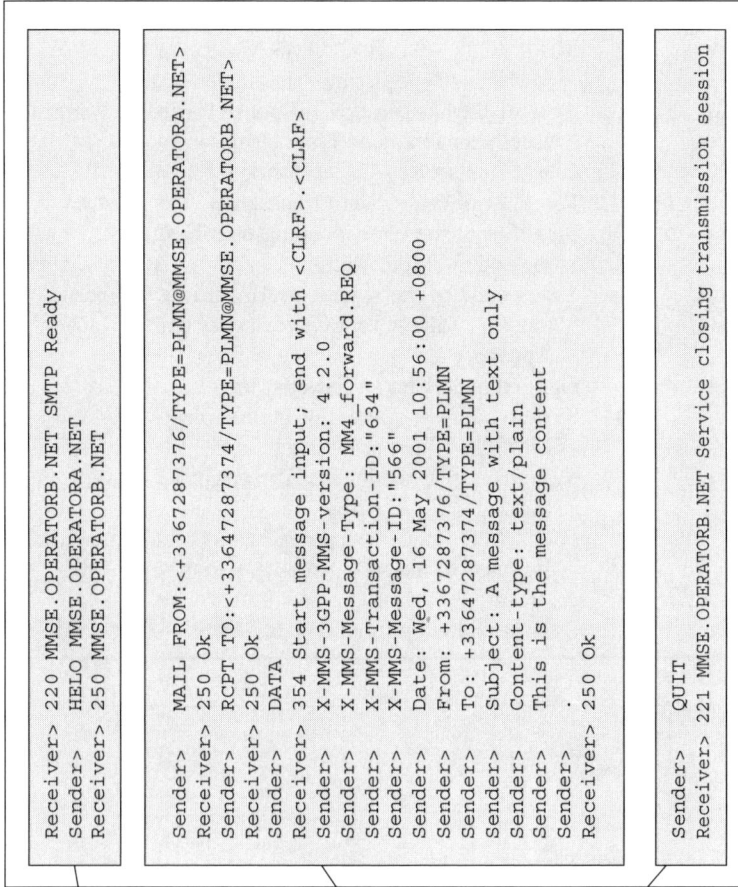

**Figure 6.30**  Example of message transfer over SMTP

**Table 6.30** MM4 read-report forward request (`MM4_read_reply_report.REQ`)

| Parameter name | Description | From 3GPP | St. |
|---|---|---|---|
| `X-Mms-3GPP-MMS-Version` | 3GPP MMS Version, MMS – Version of the MMSC Type: string – Example: `5.2.0` | Rel-4 | ● |
| `X-Mms-Message-Type` | Type of MM4 operation (request for the forward of a read report) – Type: string – Value: `MM4_read_reply_report.REQ` | Rel-4 | ● |
| `X-Mms-Transaction-ID` | Identifier of the forward transaction – Type: string | Rel-4 | ● |
| `X-Mms-Message-ID` | Identifier of the corresponding multimedia message – Type: string | Rel-4 | ● |
| `From` | Recipient address, address of the recipient of the original message – Type: string | Rel-4 | ● |
| `To` | Sender address, address of the sender of the original message – Type: string | Rel-4 | ● |
| `Date` | Message date and time, Date and time when the message was handled (read or deleted) – Type: date | Rel-4 | ● |
| `X-Mms-Ack-Request` | Request for an acknowledgement, Whether or not an acknowledgement of the forward request is requested – Values: `Yes` or `No` | Rel-4 | ○ |
| `X-Mms-Read-Status` | Message status code – Status of the corresponding multimedia message – Type: string – Values: `Read` or `Deleted without being Read`. | Rel-4 | ● |
| `X-Mms-Status-Text` | Text corresponding to the status code – Type: string | Rel-4 | ○ |
| `Sender` | System address – address to which the requested response should be sent – Type: string | Rel-4 | ● |
| `Message-ID` | Each SMTP request/response has a unique reference assigned to the `Message-ID` parameter – Type: string | Rel-4 | ● |

values assigned to `From`, `To` and `Cc` parameters are not used for routing purpose over SMTP. These values are conveyed transparently over SMTP. Consequently, these values may be formatted as email addresses or as phone numbers. Values used for routing purpose in SMTP are those assigned to `MAIL` and `RCPT` parameters.

## 6.6 MM5 Interface, MMSC–HLR

The MM5 interface enables interactions between an MMSC and other network entities such as the HLR. Operations that can be invoked over the MM5 interface include the following:

- Interrogating the HLR to obtain routing information for the purpose of forwarding a message from one MMSC to another over the MM4 interface.

**Table 6.31**  MM4 read-report forward response (`MM4_read_reply_report.RES`)

| Parameter name | Description | From 3GPP | St. |
|---|---|---|---|
| `X-Mms-3GPP-MMS-Version` | 3GPP MMS Version, MMS Version of the MMSC Type: string – Example: `5.2.0` | Rel-4 | ● |
| `X-Mms-Message-Type` | Type of MM4 operation (response for the forward of a read report) – Type: string – Value: `MM4_read_reply_report.RES` | Rel-4 | ● |
| `X-Mms-Transaction-ID` | Identifier of the forward transaction – Type: string | Rel-4 | ● |
| `X-Mms-Request-Status-Code` | Status code of the request to forward the delivery report – Values: `Ok` or error codes defined in Appendix E. | Rel-4 | ● |
| `X-Mms-Status-Text` | Optional status text – Type: string | Rel-4 | ○ |
| `Sender` | System address, address of the recipient MMSC – Type: string | Rel-4 | ● |
| `To` | System address, address of the originator MMSC – Type: string | Rel-4 | ● |
| `Message-ID` | Each SMTP request/response has a unique reference assigned to the `Message-ID` parameter – Type: string | Rel-4 | ● |
| `Date` | Date provided by the recipient MMSC – Type: string | Rel-4 | ● |

- Determination of the recipient handset's location (e.g. if the subscriber is roaming), and so on.

If the MM5 interface is present in the MMSE, then the interface is usually implemented on the basis of existing Mobile Application Part (MAP) operations. At the time of writing, no technical realization of the MM5 interface had been specified by standardization organizations.

## 6.7 MM6 Interface, MMSC–User Databases

The technical realization of the MM6 interface allows interactions between the MMSC and user databases. Unfortunately, the MM6 interface has yet to be standardized. Consequently, this interface is not covered in this book.

## 6.8 MM7 Interface, MMSC–VAS Applications

The MM7 interface enables interactions between VAS applications and an MMSC. Table 6.32 lists the eight transactions that can occur over the MM7 interface (14 PDUs).

At the time of writing, the only standardized technical realization for the MM7 interface is the one published by 3GPP in [3GPP-23.140] Release 5 (stages 2 and 3).

**Table 6.32**  List of MM7 transactions/PDUs

| Transaction | PDU name | Description | From 3GPP |
|---|---|---|---|
| Message submission | `MM7_submit.REQ` | Message submission request | Rel-5 |
| | `MM7_submit.RES` | Message submission response | |
| Message delivery | `MM7_deliver.REQ` | Message delivery request | Rel-5 |
| | `MM7_deliver.RES` | Message delivery response | |
| Cancellation | `MM7_cancel.REQ` | Message cancellation request | Rel-5 |
| | `MM7_cancel.RES` | Message cancellation response | |
| Replacement | `MM7_replace.REQ` | Message replacement request | Rel-5 |
| | `MM7_replace.RES` | Message replacement response | |
| Delivery report | `MM7_delivery_report.REQ` | Delivery-report request | Rel-5 |
| | `MM7_delivery_report.RES` | Delivery-report response | |
| Read report | `MM7_read_reply_report.REQ` | Read-report request | Rel-5 |
| | `MM7_read_reply_report.RES` | Read-report response | |
| MMSC error | `MM7_RS_error.RES` | Error indication from the MMSC to the VAS application | Rel-5 |
| VASP error | `MM7_VASP_error.RES` | Error indication from the VAS application to the MMSC | Rel-5 |

This technical realization is based on the Simple Object Access Protocol (SOAP) with HTTP at the transport layer. Figure 6.31 shows a typical network configuration allowing a VAS application to interact with several MMS clients. In this configuration, the VAS application and the MMSC play dual roles of sender and receiver of SOAP messages.

In the past, when MMS standards for the MM7 interface were not available, it was necessary for MMSC vendors to design a proprietary MM7 interface for MMSCs. Proprietary MM7 implementations rely on the transfer of proprietary commands over either SMTP or HTTP. At the time of writing, MMSC vendors were about to offer commercial MMSC solutions with support for the standardized MM7 interface as defined by 3GPP.

As for other interfaces, each transaction invoked over the MM7 interface is composed of a request and a corresponding response. HTTP-level mechanisms[4] can be

---

[4] For instance, the access authentication mechanism described in [RFC-2617] can be used for authenticating parties communicating over the MM7 interface.

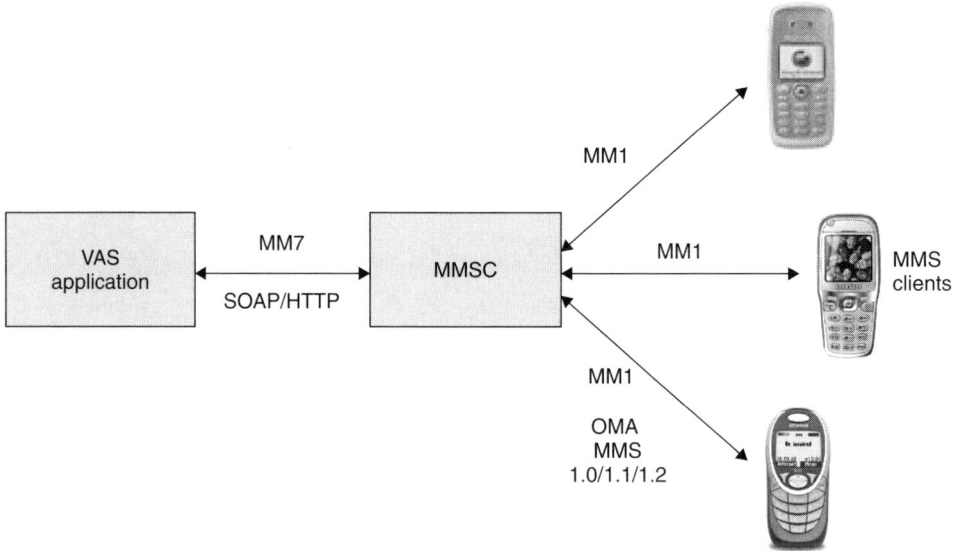

**Figure 6.31**    Network architecture with a VAS application

used in order to authenticate parties communicating over the MM7 interface. Additionally, messages over the MM7 interface can be transported over the Transport Layer Security (TLS) protocol to ensure confidentiality between communicating parties.

The support of the MM7 interface is optional for the MMSC. However, if such an interface is supported, then message submission, message delivery, transactions related to the provision of delivery reports and the management of errors are mandatory transactions to be supported by the MMSC. The support of other transactions such as message cancellation, replacement and transactions related to the management of read reports is optional for the MMSC.

The two addressing modes, phone number [ITU-E.164] and email address [RFC-2822], can be used to identify entities communicating over the MM7 interface. Communicating entities include MMS clients, VAS applications and the MMSC.

SOAP messages are structured according to SOAP technical specifications published by W3C in [W3C-SOAP] and [W3C-SOAP-ATT]. In addition, several versions of the eXtensible Markup Language (XML) schema for formatting MMS-specific SOAP messages are published by 3GPP at the following location:

```
http://www.3gpp.org/ftp/Specs/archive/23_series/23.140/schema/
```

Several versions of the schema for the MM7 interface are available and each one corresponds to a specific version of the standard [3GPP-23.140] (Release 5) as shown in Table 6.33.

**Table 6.33** XML MM7 schema name/Version of [3GPP-23.140]

| Schema name | Corresponding version of [3GPP-23.140] |
| --- | --- |
| REL-5-MM7-1-0 | v5.3.0 |
| REL-5-MM7-1-1 | v5.4.0 |
| REL-5-MM7-1-2 | v5.5.0 |
| REL-5-MM7-1-3 | v5.6.0 |

*6.8.1 Introduction to SOAP*

SOAP is a lightweight protocol for the exchange of information in distributed environments such as the MMSE. All SOAP messages are represented using XML. SOAP specifications consist of three distinct parts:

- *Envelope*: This part defines a framework for describing the content of a SOAP message and how to process it.
- *Set of encoding rules*: Encoding rules are used for expressing instances of application-defined data types.
- *Convention for representing remote procedure calls*: This convention helps entities in a distributed environment to request services from each other in an interoperable manner.

SOAP may be used over a variety of transport protocols. In the MMSE, for the realization of the MM7 interface, SOAP is used over the HTTP transport protocol. With this configuration, MM7 request messages are transferred in HTTP Post requests, whereas corresponding MM7 response messages are transferred as part of HTTP Response messages.

A SOAP message, represented using XML, consists of a SOAP envelope, a SOAP header, a SOAP body and an optional SOAP attachment as shown in Figure 6.32. For messages containing a SOAP envelope only, the media type `text/xml` is used. If the SOAP message also contains an attachment, then the media type `multipart/related` is used and the SOAP envelope is identified with the `Start` parameter of the content type. Each part of the SOAP message has, at least, the two parameters `Content-Type` and `Content-ID`.

The SOAP envelope is the first element to appear in HTTP Post requests and corresponding responses. The `SOAPAction` parameter is set to the 'Null string'. The MMSC or the VAS server is identified uniquely with a URL placed in the `host` header field of the HTTP Post method.

A request status is provided as part of the corresponding response. The status can be of three types:

The **SOAP attachment**, if present, can contain a multimedia message.

The **SOAP header** is used for transaction management.

Multimedia message
- Body part header n°1
- Body part data n°1
- Body part header n°N
- Body part data n°N

SOAP envelope

SOAP attachment

SOAP message

SOAP header

SOAP body

The **SOAP body** contains all MM7 parameters (except Transaction ID).

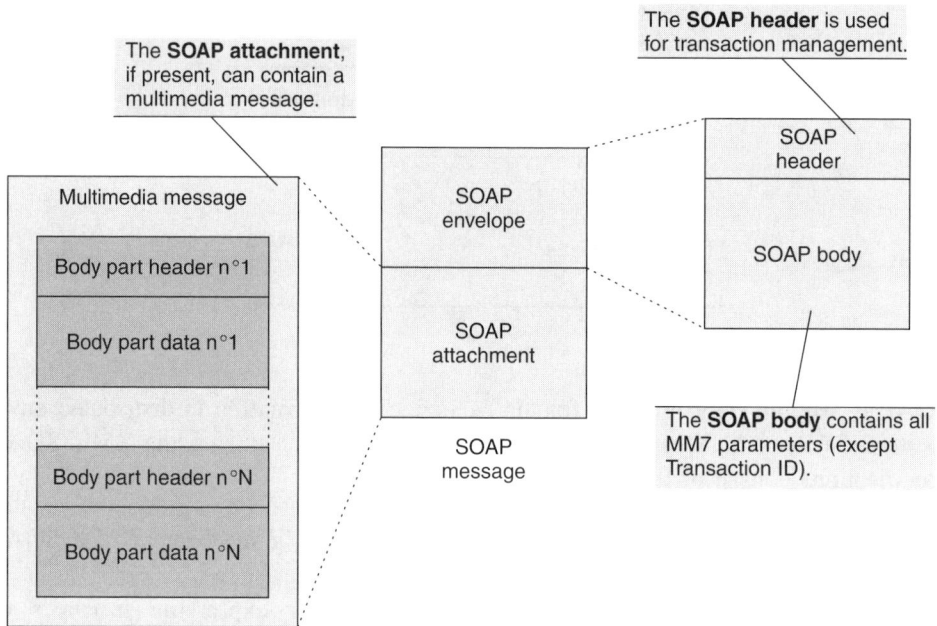

**Figure 6.32**    Structure of a SOAP message

- *Success or partial success*: This status indicates the successful or partial processing of a request. This status is composed of three parameters: StatusCode (numerical code), StatusText (human readable textual description) and Details (optional human readable detailed textual description). The classification of four-digit error identifiers to be assigned to the StatusCode is provided in Figure 6.33 and a full list of defined error codes and corresponding status texts is provided in Appendix F.
- *Processing error*: This status indicates that a fault occurred while parsing SOAP elements. These processing errors include the faultcode, faultstring and detail elements as defined in [W3C-SOAP].
- *Network error*: This status indicates that an error occurred at the HTTP level.

The 3GPP XML schema defined for the MM7 interface specifies the structure of MMS PDUs embedded in SOAP messages. The following sections describe each one of the 14 PDUs. For this purpose, XML elements composing each PDU are listed and a graphical representation of the overall PDU structure is provided. The graphical representation shows the relationships between the XML schema group elements and associated child elements. The three XML schema grouping operations 'All', 'Sequence' and 'Choice' are used for structuring MMS PDUs. The symbols in Figure 6.34 are used in this book for the graphical representations of XML structures.

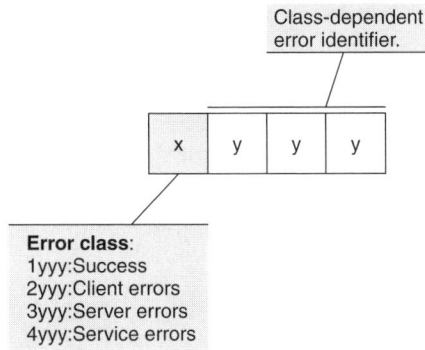

**Figure 6.33**  Classification of MM7 error codes

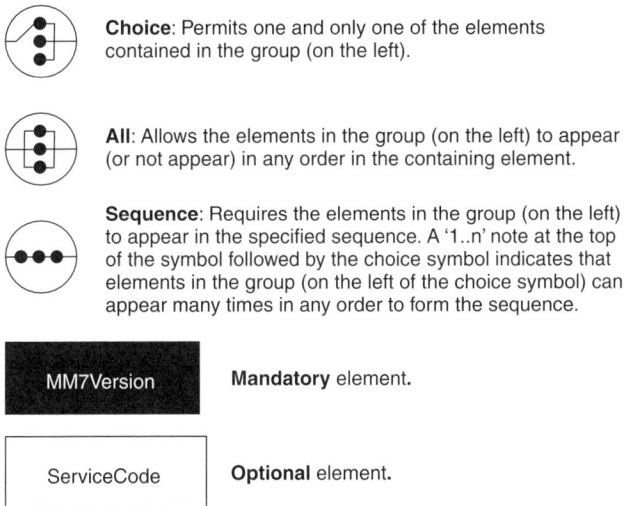

**Figure 6.34**  Graphical notations for XML schema structures

## 6.8.2 Message Submission

Regarding the MM7 interface, message submission refers to the submission of a message from an originator VAS application to an MMSC. The message is addressed to a single recipient, to multiple recipients or to a distribution list managed by the MMSC. If the MMSC accepts the submission request, then the MMSC sends back a positive response. This indicates that the submission request is accepted but does not indicate that the message has been successfully delivered to the message recipients. The transaction flow in Figure 6.35 shows interactions between the VAS application and the MMSC for the submission of a multimedia message over the MM7 interface.

In tables describing the content of requests and responses for the MM7 interface, the column named 'location' (abbreviated 'Loc.') indicates whether the corresponding parameter is placed in the SOAP header ('H'), SOAP body ('B') or SOAP attachment ('A').

The submission request `MM7_submit.REQ` is composed of parameters listed in Table 6.34. The MMSC acknowledges the submission request with the `MM7_submit.RES` response. This response is composed of parameters listed in Table 6.35.

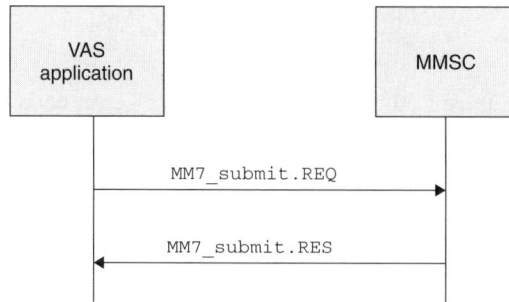

**Figure 6.35** MM7 message submission

**Table 6.34** MM7 message submission request (`MM7_submit.REQ`)

| Parameter name | Description | From 3GPP | Loc. | St. |
|---|---|---|---|---|
| TransactionID | Transaction identifier <root element of the SOAP body> | Rel-5 | H | ● |
| MessageType | Message type, Type of MM7 operation – Value: SubmitReq | Rel-5 | B | ● |
| MM7Version | Version[a] of the MM7 interface supported by the VAS application – Example: 5.6.0 | Rel-5 | B | ● |
| VASPID | <sub-element of SenderIdentification> Identification of the organization providing the VAS (VAS provider) | Rel-5 | B | ○ |
| VASID | <sub-element of SenderIdentification> Identification of the VAS | Rel-5 | B | ○ |
| SenderAddress | <sub-element of SenderIdentification> Address of the message originator | Rel-5 | B | ○ |
| Recipients | Address of the message recipient(s), Multiple recipient addresses can be specified | Rel-5 | B | ● |

**Table 6.34**   (*continued*)

| Parameter name | Description | From 3GPP | Loc. | St. |
|---|---|---|---|---|
| ServiceCode | Service code for charging purpose | Rel-5 | B | ○ |
| LinkedID | Linked identification, Linkage with another message | Rel-5 | B | ○ |
| MessageClass | Message class – Type: enumeration – Values: Informational, Advertisement or Auto | Rel-5 | B | ○ |
| TimeStamp | Date and time of message submission | Rel-5 | B | ○ |
| ExpiryDate | Message time of expiry | Rel-5 | B | ○ |
| EarliestDeliveryTime | Message earliest time of delivery | Rel-5 | B | ○ |
| DeliveryReport | Request for a delivery report – Values: True or False | Rel-5 | B | ○ |
| ReadReply | Request for a read report – Values: True or False | Rel-5 | B | ○ |
| ReplyCharging | Request for reply charging – Parameter has two attributes: replyDeadline and replyChargingSize – Presence of this parameter means that reply charging is requested | Rel-5 | B | ○ |
| replyDeadline | <attribute of ReplyCharging> Reply charging deadline – Type: absolute or relative date format | Rel-5 | B | ○ |
| replyChargingSize | <attribute of ReplyCharging> Reply charging max. message size | Rel-5 | B | ○ |
| Priority | Message priority – Values: High, Normal or Low | Rel-5 | B | ○ |
| Subject | Message subject | Rel-5 | B | ○ |
| DistributionIndicator | Whether or not the message can be redistributed freely – Values: True or False | Rel-5 | B | ○ |
| ChargedParty | Indication of the party(ies) that should be charged for the cost of handling the message – Values: Sender, Recipient, Both or Neither | Rel-5 | B | ○ |
| Content | Message content | Rel-5 | B | ○ |
| allowAdaptations | <attribute of Content> Whether or not content adaptations are allowed – Values: True or False | Rel-5 | B | ○ |
| Content-Type | Message content type | Rel-5 | A | ● |

[a]From Release 6, the MMS version assigned to this parameter is the version of the specification for which the XML MM7 schema has changed most recently. This avoids having to produce a new schema each time [3GPP-23.140] has evolved without schema updates.

**Table 6.35**  MM7 message submission response (`MM7_submit.RES`)

| Parameter name | Description | From 3GPP | Loc. | St. |
|---|---|---|---|---|
| `TransactionID` | Transaction identifier <root element of the SOAP body> | Rel-5 | H | ● |
| `MessageType` | Message type, Type of MM7 operation<br>Value: `SubmitRsp` | Rel-5 | B | ● |
| `MM7Version` | Version of the MM7 interface supported by the MMSC<br>Example: `5.6.0` | Rel-5 | B | ● |
| `MessageID` | Message identifier generated by the MMSC<br>Condition: available only if the MMSC accepts the submission request | Rel-5 | B | © |
| `StatusCode` | <sub-element of `Status`><br>Status of the corresponding request completion. | Rel-5 | B | ● |
| `StatusText` | <sub-element of `Status`><br>Textual description of the status of the request completion. | Rel-5 | B | ○ |
| `Details` | <sub-element of `Status`><br>Human readable detailed textual description of the corresponding request status. | Rel-5 | B | ○ |

Figure 6.36 shows a graphical representation of the message submission request body, whereas Figure 6.37 shows a graphical representation of the message submission response body.

Figure 6.38 shows an example of submission request embedded in an HTTP Post request, whereas Figure 6.39 shows the corresponding response.

### 6.8.3 Message Delivery

Regarding the MM7 interface, message delivery refers to the delivery of a multimedia message from the MMSC to a VAS application. The MMSC may deliver the message to the VAS application along with a linked identification. This identification can be conveyed as part of a subsequent message submission from the VAS application to indicate that the submitted message is related to a previously delivered message. The transaction flow in Figure 6.40 shows interactions between the MMSC and the VAS application for the delivery of a multimedia message over the MM7 interface.

The delivery request `MM7_deliver.REQ` is composed of parameters listed in Table 6.36. Figure 6.41 shows a graphical representation of the message delivery request body.

The VAS application acknowledges the delivery request with the `MM7_deliver.RES` response. This response is composed of parameters listed in Table 6.37.

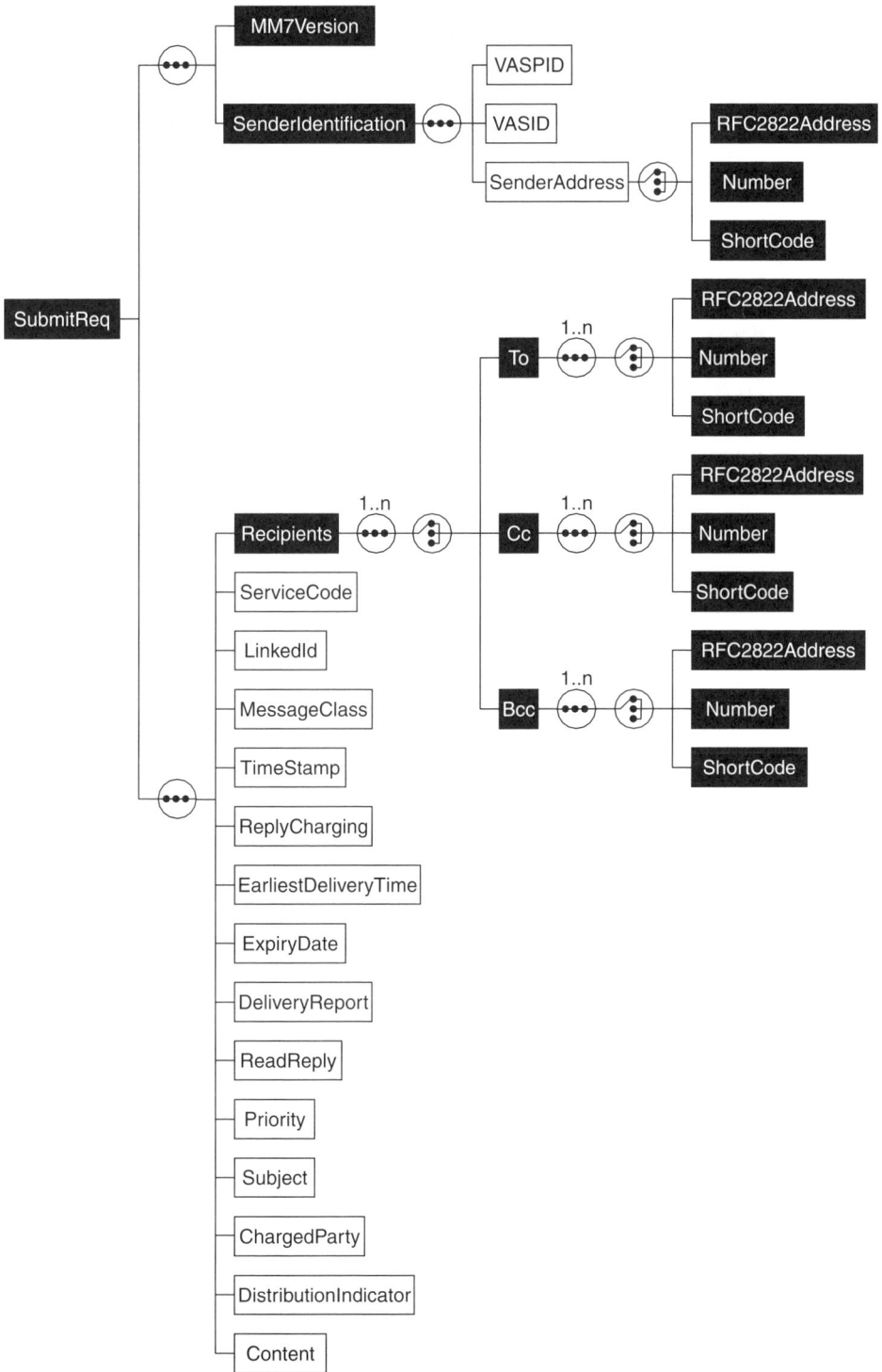

**Figure 6.36**  Graphical representation of the submission request PDU body

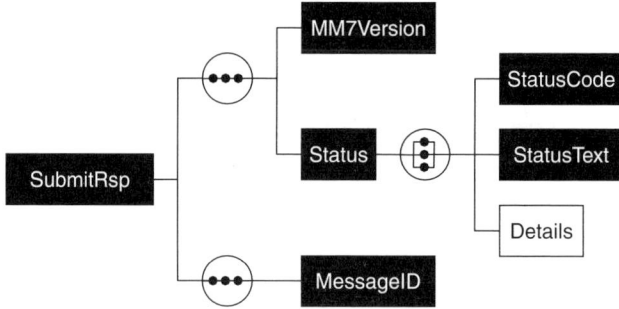

**Figure 6.37** Graphical representation of the submission response PDU body

**Table 6.36** MM7 message delivery request (`MM7_deliver.REQ`)

| Parameter name | Description | From 3GPP | Loc. | St. |
|---|---|---|---|---|
| TransactionID | Transaction identifier <root element of the SOAP body> | Rel-5 | H | ● |
| MessageType | Message type, Type of MM7 operation Value: DeliverReq | Rel-5 | B | ● |
| MM7Version | Version of the MM7 interface supported by the MMSC Example: 5.6.0 | Rel-5 | B | ● |
| MMSRelayServerID | Identification of the MMSC | Rel-5 | B | ○ |
| Sender | Address of the message sender | Rel-5 | B | ● |
| Recipients | Address of the message recipient(s), Multiple recipient addresses can be specified | Rel-5 | B | ○ |
| LinkedID | Linked identification, Linkage with another message | Rel-5 | B | ○ |
| TimeStamp | Date and time of message submission | Rel-5 | B | ○ |
| ReplyChargingID | Identification of the reply charging transaction associated to the reply message | Rel-5 | B | ○ |
| Priority | Message priority – Values: High, Normal or Low | Rel-5 | B | ○ |
| Subject | Message subject | Rel-5 | B | ○ |
| Content | Message content | Rel-5 | B | ○ |
| Content-Type | Message content – type | Rel-5 | A | ● |

Figure 6.42 shows a graphical representation of the message delivery response body.

### 6.8.4 Message Cancellation

In the context of value-added services, message cancellation refers to the possibility for a VAS application to cancel the delivery of a multimedia message. Upon receipt of such a cancel request, the MMSC cancels the delivery of the associated message

**Table 6.37**  MM7 message delivery response (`MM7_deliver.RES`)

| Parameter name | Description | From 3GPP | Loc. | St. |
|---|---|---|---|---|
| `TransactionID` | Transaction identifier <root element of the SOAP body> | Rel-5 | H | ● |
| `MessageType` | Message type, Type of MM7 operation Value: `DeliverRsp` | Rel-5 | B | ● |
| `MM7Version` | Version of the MM7 interface supported by the VAS application – Example: `5.6.0` | Rel-5 | B | ● |
| `ServiceCode` | Service code for charging purpose | Rel-5 | B | ○ |
| `StatusCode` | <sub-element of `Status`> Status of the corresponding request completion. | Rel-5 | B | ● |
| `StatusText` | <sub-element of `Status`> Textual description of the status of the request completion. | Rel-5 | B | ○ |
| `Details` | <sub-element of `Status`> Human readable detailed textual description of the corresponding request status. | Rel-5 | B | ○ |

to all message recipients to which the associated notification has not yet been sent out. The transaction flow in Figure 6.43 shows interactions over the MM7 interface between the MMSC and the VAS application for the cancellation of the delivery of a multimedia message.

The cancellation request `MM7_cancel.REQ` is composed of parameters listed in Table 6.38. Figure 6.44 shows a graphical representation of the message cancel request body.

The MMSC acknowledges the cancellation request with the `MM7_cancel.RES` response. This response is composed of parameters listed in Table 6.39. Figure 6.45 shows a graphical representation of the message cancellation response body.

### 6.8.5 Message Replacement

In the context of value-added services, message replacement refers to the possibility for a VAS application to replace a previously submitted multimedia message prior to its delivery. Upon receipt of such a replacement request, the MMSC replaces the previously submitted message with the new message specified as part of the replacement request. Only messages that have not yet been retrieved or forwarded can be replaced. The replacement request contains a number of parameters that overwrite those associated with the previously submitted message. If a parameter was associated with the previously submitted message but is not provided as part of the replacement request, then this parameter is retained for the new message. The transaction flow in

```
POST /mmsc/mm7 HTTP/1.1
Host: mms.operator.com

Content-Type: Multipart/Related; boundary="PrimaryBoundary"; type=text/xml;
    start="<id_start>"
Content-Length: nnnn
SOAPAction: ""

--PrimaryBoundary
Content-Type: text/xml; charset="utf-8"
Content-ID: <id_start>

<?xml version='1.0' ?>
<env:Envelope xmlns:env="http://schemas.xmlsoap.org/soap/envelope/">

        <env:Header>
            <mm7:TransactionID xmlns="http://www.3gpp.org/ftp/Specs/archive/23_series/23.140/schema/REL-5-MM7-1-3"
             env:mustUnderstand="1" />
                    vasp0324-sub
            </mm7:TransactionID>
        </env:Header>

        <env:Body>
            <mm7:SubmitReq xmlns:mm7="http://www.3gpp.org/ftp/Specs/archive/23_series/23.140/schema/REL-5-MM7-1-3">
                    <MM7Version>5.6.0</MM7Version>
                    <SenderIdentification>
                            <VASPID>w-up</VASPID>
                            <VASID>Weather</VASID>
                    </SenderIdentification>
                    <Recipients>
                            <To>
                                <Number>0147287355</Number>
                                <RFC2822Address>0147287366@operator.com</RFC2822Address>
                            </To>
                            <Cc>
                                <Number>0147243477</Number>
                            </Cc>
                            <Bcc>
                                <Number>0147243488</Number>
                            </Bcc>
                    </Recipients>
                    <ServiceCode>weather-france</ServiceCode>
                    <LinkID>mms00000006</LinkID>
                    <MessageClass>Informational</MessageClass>
                    <TimeStamp>2002-05-05T09:30:45-01:00</TimeStamp>
                    <EarliestDeliveryTime>2002-05-05T09:30:45-01:00</EarliestDeliveryTime>
                    <ExpiryDate>P5D</ExpiryDate>
                    <DeliveryReport>True</DeliveryReport>
                    <Priority>Normal</Priority>
                    <Subject>Weather for today</Subject>
                    <Content href="cid:weatherpic@w-up.com" ; allowAdptations="true" />
            </mm7:SubmitReq>
        </env:Body>
</env:Envelope>

--PrimaryBoundary
Content-Type: multipart/mixed; boundary="SecondaryBoundary"
Content-ID: <weatherpic@w-up.com>

--SecondaryBoundary
Content-Type: text/plain; charset="us-ascii"

The weather picture.

--SecondaryBoundary
Content-Type: image/gif
Content-ID: <weather.gif>
Content-Transfer-Encoding: base64

R0IGODfg...

--SecondaryBoundary--
--PrimaryBoundary--
```

(Left margin labels: SOAP envelope, SOAP attachment)

**Figure 6.38**   Example of message submission request over the MM7 interface

Figure 6.46 shows interactions over the MM7 interface between the MMSC and the VAS application for the replacement of a previously submitted multimedia message.

The replacement request `MM7_replace.REQ` is composed of parameters listed in Table 6.40. Figure 6.47 shows a graphical representation of the message replacement request body.

```
HTTP/1.1 200 OK
Content-Type: text/xml; charset="utf8"
Content-Length: nnnn

<?xml version='1.0' ?>
<env:Envelope xmlns:env="http://schemas.xmlsoap.org/soap/soap-envelope">
    <env:Header>
        <mm7:TransactionID xmlns:mm7="http://www.3gpp.org/ftp/Specs/archive/23_series/23.140/schema/REL-5-MM7-1-3"
            env:mustUnderstand="1">
            vasp0324-sub
        </mm7:TransactionID>
    </env:Header>
    <env:Body>
        <mm7:SubmitRsp xmlns:mm7="http://www.3gpp.org/ftp/Specs/archive/23_series/23.140/schema/REL-5-MM7-1-3">
            <MM7Version>5.6.0</MM7Version>
            <Status>
                <StatusCode>1000</StatusCode>
                <StatusText>Message sent</StatusText>
                <MessageID>123456</MessageID>
            </Status>
        </mm7:SubmitRsp>
    </env:Body>
</env:Envelope>
```

**Figure 6.39** Example of message submission response over the MM7 interface

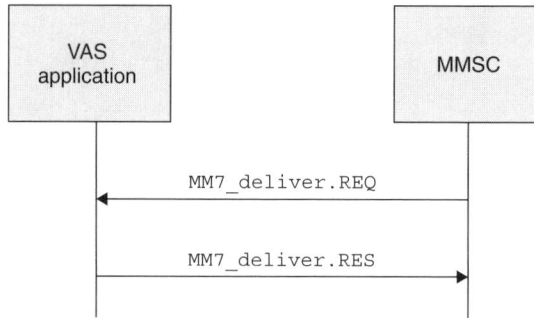

**Figure 6.40** MM7 message delivery

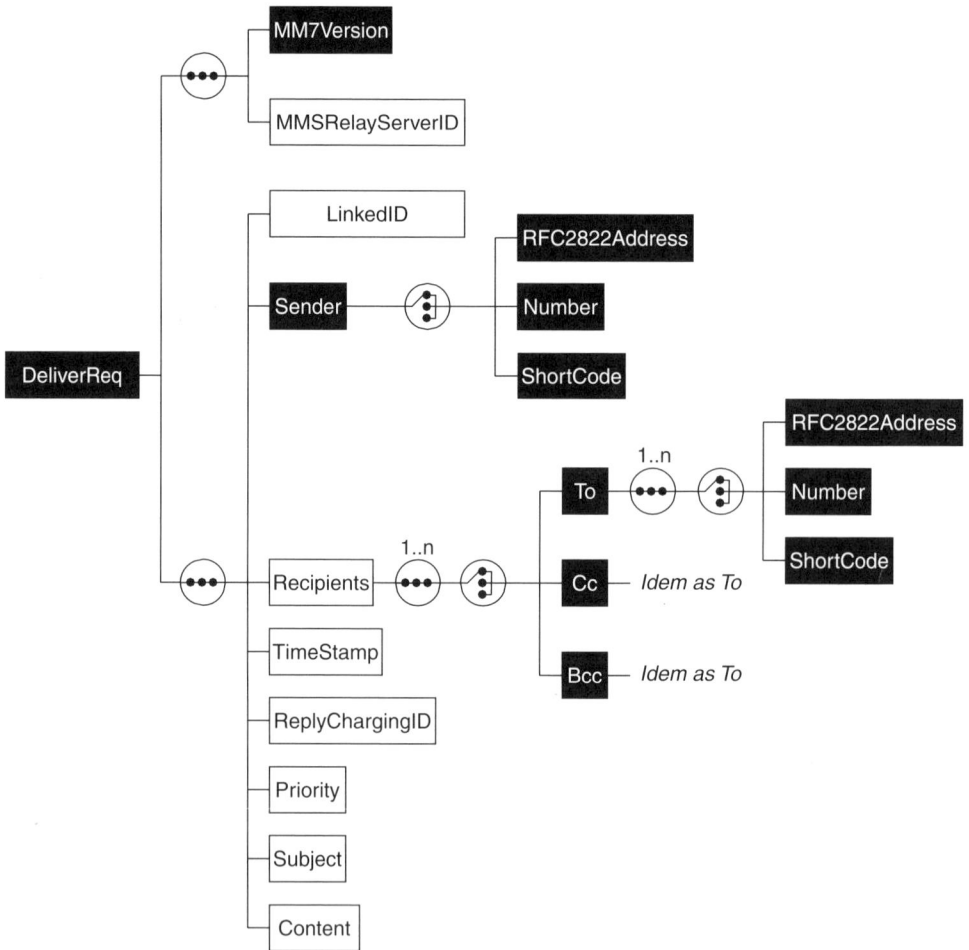

**Figure 6.41** Graphical representation of the delivery request PDU body

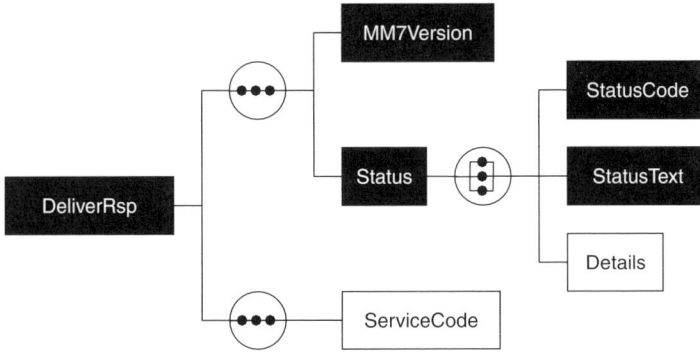

**Figure 6.42**  Graphical representation of the delivery response PDU body

**Table 6.38**  MM7 message cancel request (`MM7_cancel.REQ`)

| Parameter name | Description | From 3GPP | Loc. | St. |
|---|---|---|---|---|
| TransactionID | Transaction identifier <root element of the SOAP body> | Rel-5 | H | ● |
| MessageType | Message type, Type of MM7 operation Value: `CancelReq` | Rel-5 | B | ● |
| MM7Version | Version of the MM7 interface supported by the VAS application – Example: `5.6.0` | Rel-5 | B | ● |
| VASPID | <sub-element of `SenderIdentification`> Identification of the organization providing the value added service (VASP) | Rel-5 | B | ○ |
| VASID | <sub-element of `SenderIdentification`> Identification of the VAS | Rel-5 | B | ○ |
| SenderAddress | <sub-element of `SenderIdentification`> Address of the message originator | Rel-5 | B | ○ |
| MessageID | Identifier of the message for which the delivery is to be cancelled. This identifier was provided by the MMSC in the response of the associated submission request | Rel-5 | B | ● |

The MMSC acknowledges the replacement request with the `MM7_replace.RES` response. This response is composed of parameters listed in Table 6.41.

The structure of the replacement response body is similar to the one of the cancellation response body. This structure is shown in Figure 6.45.

### 6.8.6 Delivery Report

In the context of value-added services, a VAS application has the ability to request, as part of a message submission request, the generation of a delivery report. If allowed by

**Table 6.39**   MM7 message cancellation response (`MM7_cancel.RES`)

| Parameter name | Description | From 3GPP | Loc. | St. |
|---|---|---|---|---|
| TransactionID | Transaction identifier <root element of the SOAP body> | Rel-5 | H | ● |
| MessageType | Message type, Type of MM7 operation Value: `CancelRsp` | Rel-5 | B | ● |
| MM7Version | Version of the MM7 interface supported by the VAS application – Example: `5.6.0` | Rel-5 | B | ● |
| StatusCode | <sub-element of `Status`> Status of the corresponding request completion. | Rel-5 | B | ● |
| StatusText | <sub-element of `Status`> Textual description of the status of the request completion. | Rel-5 | B | ○ |
| Details | <sub-element of `Status`> Human readable detailed textual description of the corresponding request status. | Rel-5 | B | ○ |

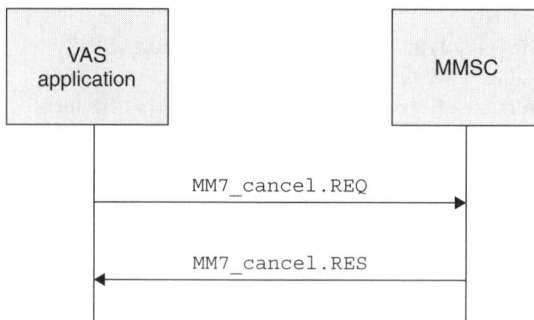

**Figure 6.43**   MM7 message cancellation

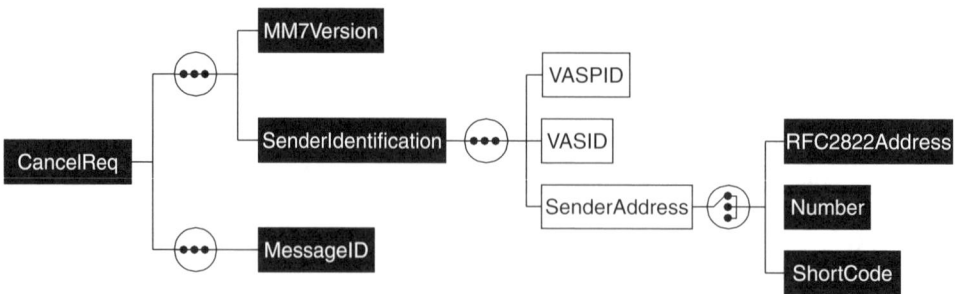

**Figure 6.44**   Graphical representation of the cancel request PDU body

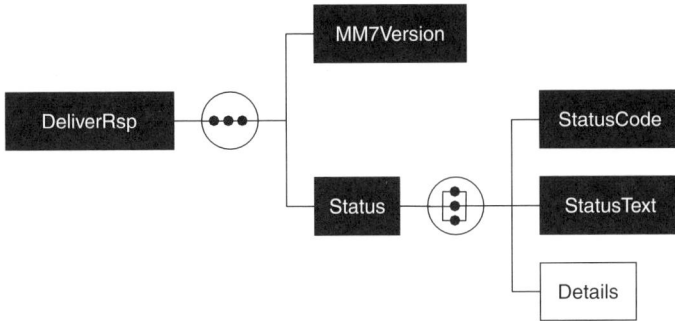

**Figure 6.45**   Graphical representation of the cancel/replace response PDU body

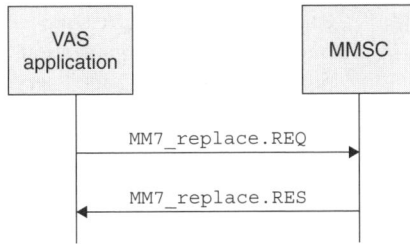

**Figure 6.46**   MM7 message replacement

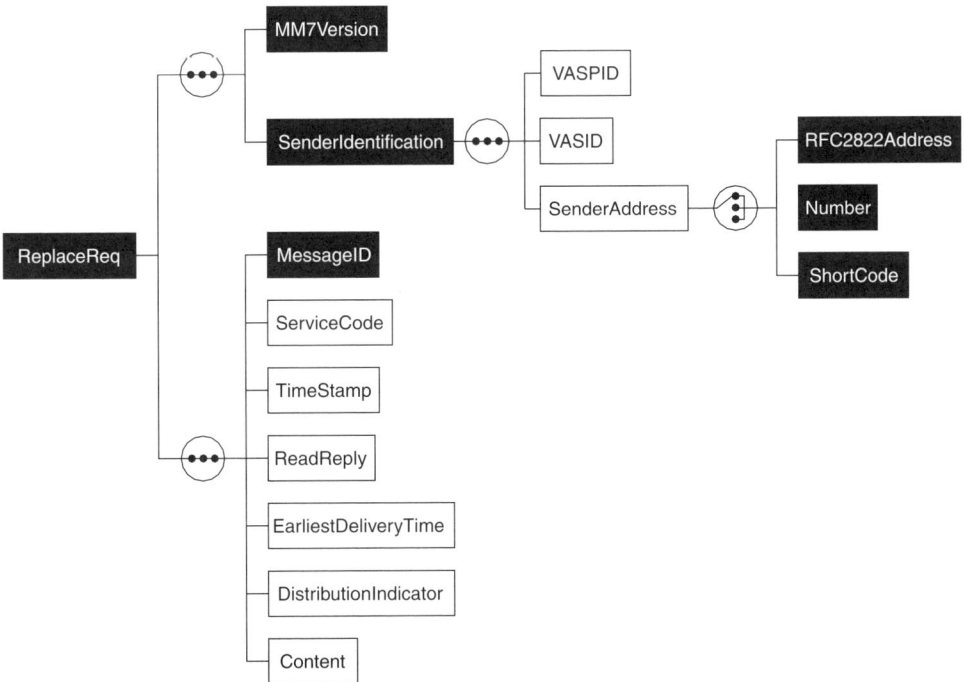

**Figure 6.47**   Graphical representation of the replace request PDU body

**Table 6.40**   MM7 message replace request (`MM7_replace.REQ`)

| Parameter name | Description | From 3GPP | Loc. | St. |
|---|---|---|---|---|
| TransactionID | Transaction identifier <root element of the SOAP body> | Rel-5 | H | ● |
| MessageType | Message type, Type of MM7 operation<br>Value: `ReplaceReq` | Rel-5 | B | ● |
| MM7Version | Version of the MM7 interface supported by the VAS application – Example: `5.6.0` | Rel-5 | B | ● |
| VASPID | <sub-element of `SenderIdentification`><br>Identification of the organization providing the value added service (VASP) | Rel-5 | B | ○ |
| VASID | <sub-element of `SenderIdentification`><br>Identification of the value-added service | Rel-5 | B | ○ |
| SenderAddress | <sub-element of `SenderIdentification`><br>Address of the message originator | Rel-5 | B | ○ |
| MessageID | Identification of the message to be replaced. This identification was provided by the MMSC in the response of the associated submission request | Rel-5 | B | ● |
| ServiceCode | Service code for charging purpose | Rel-5 | B | ○ |
| TimeStamp | Date and time of message submission | Rel-5 | B | ○ |
| EarliestDeliveryTime | Message earliest time of delivery | Rel-5 | B | ○ |
| ReadReply | Request for a read report – Values: `True` or `False` | Rel-5 | B | ○ |
| allowAdaptations | <attribute of `Content`><br>Whether or not content adaptations are allowed – Values: `True` or `False` | Rel-5 | B | ○ |
| DistributionIndicator | Whether or not the message can be redistributed freely<br>Values: `True` or `False` | Rel-5 | B | ○ |
| Content | Message content | Rel-5 | B | ○ |
| Content-Type | Message content type<br>Condition: if a message content is available then the content type must also be present in the PDU. | Rel-5 | A | © |

**Table 6.41**   MM7 message replace response (`MM7_replace.RES`)

| Parameter name | Description | From 3GPP | Loc. | St. |
|---|---|---|---|---|
| TransactionID | Transaction identifier <root element of the SOAP body> | Rel-5 | H | ● |
| MessageType | Message type, Type of MM7 operation Value: `ReplaceRsp` | Rel-5 | B | ● |
| MM7Version | Version of the MM7 interface supported by the VAS application – Example: `5.6.0` | Rel-5 | B | ● |
| StatusCode | <sub-element of `Status`> Status of the corresponding request completion. | Rel-5 | B | ● |
| StatusText | <sub-element of `Status`> Textual description of the status of the request completion. | Rel-5 | B | ○ |
| Details | <sub-element of `Status`> Human readable detailed textual description of the corresponding request status. | Rel-5 | B | ○ |

**Table 6.42**   MM7 message delivery report request (`MM7_delivery_report.REQ`)

| Parameter name | Description | From 3GPP | Loc. | St. |
|---|---|---|---|---|
| TransactionID | Transaction identifier <root element of the SOAP body> | Rel-5 | H | ● |
| MessageType | Message type, Type of MM7 operation Value: `DeliveryReportReq` | Rel-5 | B | ● |
| MM7Version | Version of the MM7 interface supported by the MMSC Example: `5.6.0` | Rel-5 | B | ● |
| MMSRelayServerID | Identification of the MMSC | Rel-5 | B | ○ |
| MessageID | Identification of the message to which the delivery report relates | Rel-5 | B | ● |
| Recipient | Address of the message recipient | Rel-5 | B | ● |
| Sender | Address of the VAS application which previously submitted the message | Rel-5 | B | ● |
| TimeStamp | Date and time of associated message was handled | Rel-5 | B | ○ |
| MMStatus | Delivery status of the associated message – Values: `Expired`, `Retrieved`, `Rejected`, `Indeterminate` or `Forwarded` | Rel-5 | B | ● |
| StatusText | Textual description of the status of the request completion. | Rel-5 | B | ○ |

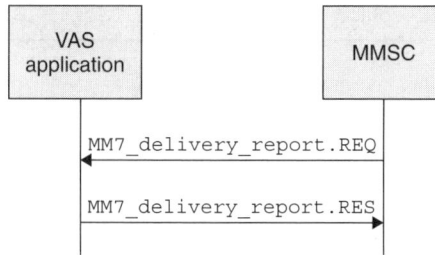

**Figure 6.48**   MM7 delivery report

**Table 6.43**   MM7 delivery report response (`MM7_delivery_report.RES`)

| Parameter name | Description | From 3GPP | Loc. | St. |
|---|---|---|---|---|
| TransactionID | Transaction identifier <root element of the SOAP body> | Rel-5 | H | ● |
| MessageType | Message type, Type of MM7 operation Value: `DeliveryReportRsp` | Rel-5 | B | ● |
| MM7Version | Version of the MM7 interface supported by the VAS application – Example: `5.6.0` | Rel-5 | B | ● |
| StatusCode | <sub-element of `Status`> Status of the corresponding request completion. | Rel-5 | B | ● |
| StatusText | <sub-element of `Status`> Textual description of the status of the request completion. | Rel-5 | B | ○ |
| Details | <sub-element of `Status`> Human readable detailed textual description of the corresponding request status. | Rel-5 | B | ○ |

the message recipient, the MMSC generates a delivery report upon message retrieval, forwarding, deletion, rejection, and so on. Note that if the message was submitted to multiple recipients, then several delivery reports may be received by the originator VAS application. The transaction flow in Figure 6.48 shows interactions between the MMSC and the VAS application for the transfer of a delivery report over the MM7 interface.

The request for providing a delivery report `MM7_delivery_report.REQ` is composed of parameters listed in Table 6.42. Figure 6.49 shows a graphical representation of the delivery report request body.

The MMSC acknowledges the request with the `MM7_delivery_report.RES` response. This response is composed of parameters listed in Table 6.43. Figure 6.50 shows a graphical representation of the delivery report response body.

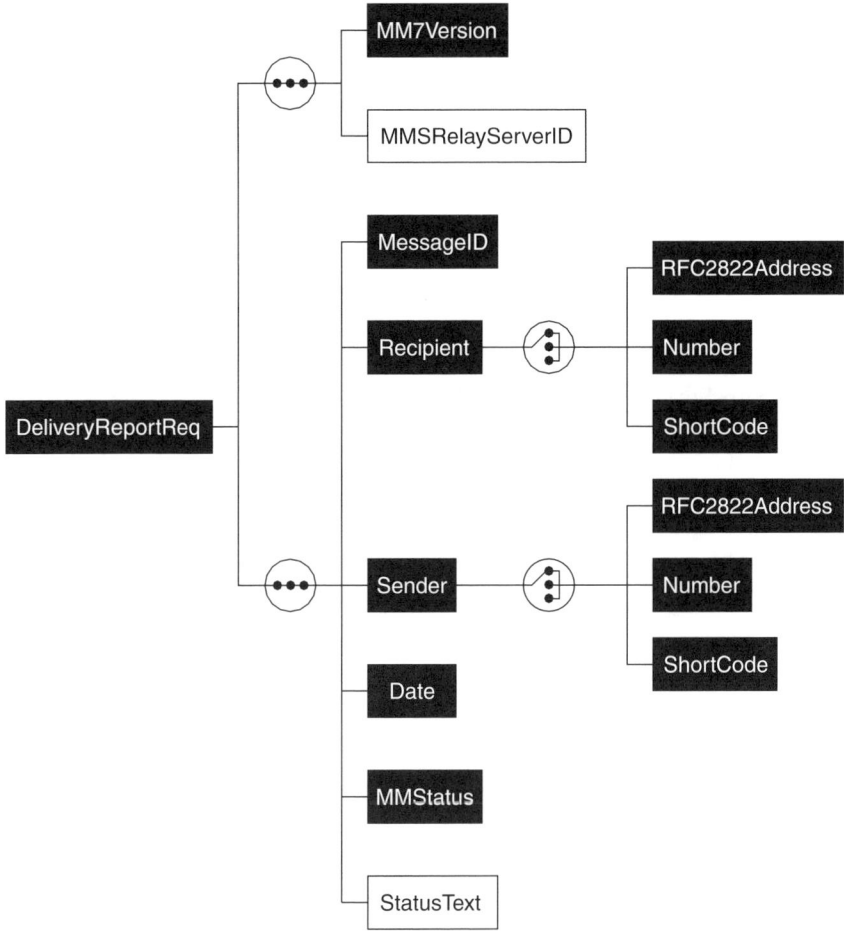

**Figure 6.49** Graphical representation of the delivery/read report request PDU body

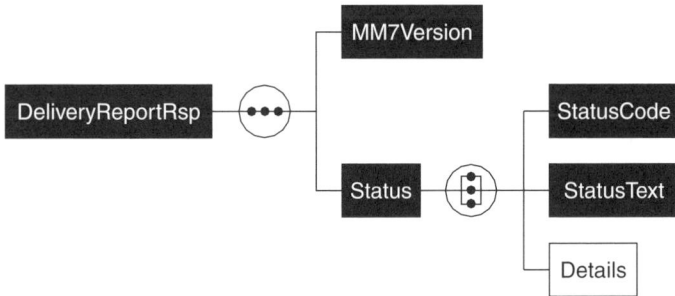

**Figure 6.50** Graphical representation of the delivery/read report response PDU body

## 6.8.7 Read Report

In the context of value-added services, a VAS application has the ability to request, as part of a message submission request, the generation of a read report. If allowed by the message recipient, the recipient MMS client generates a read report upon message reading, deletion, and so on. Note that, if the message was submitted to multiple recipients, then several read reports may be received by the originator VAS application. The transaction flow in Figure 6.51 shows interactions between the MMSC and the VAS application for the transfer of a read report over the MM7 interface.

The request for providing a read-report `MM7_read_reply_report.REQ` is composed of parameters listed in Table 6.44.

The MMSC acknowledges the request with the `MM7_read_reply_report.RES` response. This response is composed of parameters listed in Table 6.45.

**Table 6.44** MM7 message read-report request (`MM7_read_reply_report.REQ`)

| Parameter name | Description | From 3GPP | Loc. | St. |
|---|---|---|---|---|
| TransactionID | Transaction identifier <root element of the SOAP body> | Rel-5 | H | ● |
| MessageType | Message type, Type of MM7 operation Value: `ReadReplyReq` | Rel-5 | B | ● |
| MM7Version | Version of the MM7 interface supported by the MMSC – Example: `5.6.0` | Rel-5 | B | ● |
| MMSRelayServerID | Identification of the MMSC | Rel-5 | B | ○ |
| MessageID | Identification of the message to which the read report relates | Rel-5 | B | ● |
| Recipient | Address of the message recipient | Rel-5 | B | ● |
| Sender | Address of the VAS application which previously submitted the message | Rel-5 | B | ● |
| TimeStamp | Date and time the associated message was handled | Rel-5 | B | ○ |
| MMStatus | Delivery status of the associated message – Values: `Read`, `Deleted` or `Indeterminate` | Rel-5 | B | ● |
| StatusText | Textual description of the status of the request completion. | Rel-5 | B | ○ |

**Table 6.45** MM7 read-report response (`MM7_read_reply_report.RES`)

| Parameter name | Description | From 3GPP | Loc. | St. |
|---|---|---|---|---|
| TransactionID | Transaction identifier <root element of the SOAP body> | Rel-5 | H | ● |
| MessageType | Message type, Type of MM7 operation<br>Value: ReadReplyRsp | Rel-5 | B | ● |
| MM7Version | Version of the MM7 interface supported by the VAS application – Example: 5.6.0 | Rel-5 | B | ● |
| StatusCode | <sub-element of Status><br>Status of the corresponding request completion. | Rel-5 | B | ● |
| StatusText | <sub-element of Status><br>Textual description of the status of the request completion. | Rel-5 | B | ○ |
| Details | <sub-element of Status><br>Human readable detailed textual description of the corresponding request status. | Rel-5 | B | ○ |

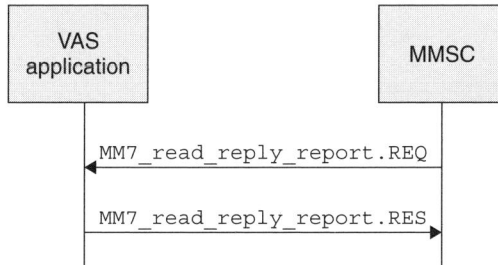

**Figure 6.51** MM7 read report

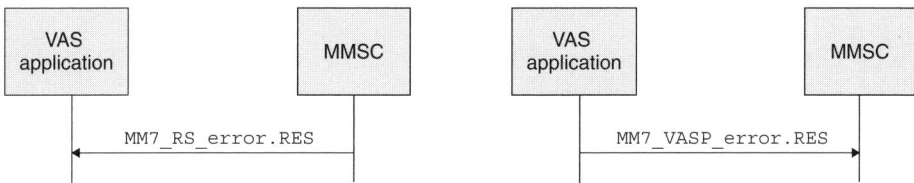

**Figure 6.52** MM7 generic errors

**Table 6.46**   MM7 MMSC error response (`MM7_RS_error.RES`)

| Parameter name | Description | From 3GPP | Loc. | St. |
|---|---|---|---|---|
| TransactionID | Transaction identifier <root element of the SOAP body> | Rel-5 | H | ● |
| MessageType | Message type, Type of MM7 operation Value: `RSErrorRsp` | Rel-5 | B | ● |
| MM7Version | Version of the MM7 interface supported by the VAS application – Example: `5.6.0` | Rel-5 | B | ● |
| StatusCode | <sub-element of `Status`> Status of the corresponding request completion. | Rel-5 | B | ● |
| StatusText | <sub-element of `Status`> Textual description of the status of the request completion. | Rel-5 | B | ○ |
| Details | <sub-element of `Status`> Human readable detailed textual description of the corresponding request status. | Rel-5 | B | ○ |

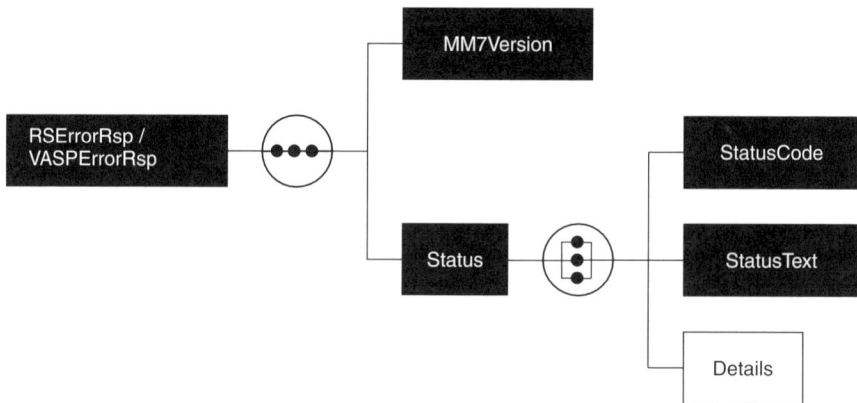

**Figure 6.53**   Graphical representation of the generic error response PDU bodies

Body structures of the read-report request and response are similar to those of the delivery-report request and response. Consequently, graphical representations of read-report request and response body structures are respectively shown in Figure 6.49 and Figure 6.50.

### 6.8.8 Generic Error Handling

In the situation in which the MMSC or the VAS application receives a request that it cannot process, a generic error notification can be used. The generic error notification

**Table 6.47** MM7 VAS application error response (`MM7_VASP_error.RES`)

| Parameter name | Description | From 3GPP | Loc. | St. |
|---|---|---|---|---|
| `TransactionID` | Transaction identifier <root element of the SOAP body> | Rel-5 | H | ● |
| `MessageType` | Message type, Type of MM7 operation<br>Value: `VASPErrorRsp` | Rel-5 | B | ● |
| `MM7Version` | Version of the MM7 interface supported by the VAS application – Example: `5.6.0` | Rel-5 | B | ● |
| `StatusCode` | <sub-element of `Status`><br>Status of the corresponding request completion. | Rel-5 | B | ● |
| `StatusText` | <sub-element of `Status`><br>Textual description of the status of the request completion. | Rel-5 | B | ○ |
| `Details` | <sub-element of `Status`><br>Human readable detailed textual description of the corresponding request status. | Rel-5 | B | ○ |

always contains the identification of the corresponding request (value assigned to the `TransactionID` parameter) to which it relates. Two generic error notifications can be used as shown in the transaction flows in Figure 6.52.

The MMSC notifies a VAS application of a generic error with the `MM7_RS_error.RES` response. This response is composed of parameters listed in Table 6.46.

The VAS application notifies the MMSC of a generic error with the `MM7_VASP_error.RES` response. This response is composed of parameters listed in Table 6.47.

Bodies of the two generic error responses have the same structure as graphically represented in Figure 6.53.

## 6.9 MM8 Interface, MMSC–Billing System

The MM8 interface interconnects the MMSC and the billing system. 3GPP has standardized charging data records for MMS (see Section 4.13), but a mechanism for conveying these charging data records from the MMSC to the billing system is still to be standardized. Consequently, network operators use existing standard transport protocols or proprietary transport protocols for this purpose.

# 7

# Standard Compliance and Interoperability

Once Multimedia Messaging Service (MMS) standards have been published, vendors can develop devices on the basis of these standards. Of course, devices (MMSC and MMS clients) are implemented with more or less features to match specific market requirements. However, devices claiming conformance to the same standards are expected to interoperate efficiently. This is the prime objective of a standard and a key enabler for the success of any communication service.

This chapter introduces the concepts of standard conformance and interoperability testing. It also describes formal methods that have been set by the Open Mobile Alliance (OMA) in order to test interoperability between devices prior to their market launch. The chapter also describes mechanisms that allow devices complying with different versions of the same standard to interoperate in a satisfactory manner.

## 7.1 Standard Conformance and Interoperability Testing

Standard conformance refers to the adherence of a device implementation to the normative requirements of one or more technical specifications (standards) and interoperability testing (IOT) refers to the process of testing that two or more devices interoperate properly and are in line with the normative requirements.

This section focuses on the formal methods for standard conformance and interoperability testing in the scope of the OMA process. The Third Generation Partnership Project (3GPP) focuses on radio-level conformance testing and has no intention so far of providing a framework for interoperability testing at the application/service level.

A formal certification programme for MMS does not exist as yet, but one may be elaborated in the near future.

### 7.1.1 Static Conformance Requirements

Each OMA specification can include a Static Conformance Requirement (SCR) definition (appendix section of the standard). An SCR definition for a specification provides

*Multimedia Messaging Service: An Engineering Approach to MMS*   Gwenaël Le Bodic
© 2003 John Wiley & Sons, Ltd   ISBN: 0-470-86253-X

| Item | Function | Reference | Status | Requirement |
|------|----------|-----------|--------|-------------|
| MMSCTR-NTF-S-001 | Notification Transcation between MMS Proxy-Relay and Receiving MMS Client | 7.2 | M | MMSCTR-NTF-S-002 AND MMSCTR-NTF-S-003 |
| MMSCTR-NTF-S-002 | MMS Proxy-Relay Sending M-Notification.ind to Receiving MMS Client | 7.2.1 | O | MMSE-S-077 AND MMSCTR-PSH-S-002 |
| MMSCTR-NTF-S-003 | Receiving MMS Client Sending M-NotifyResp.ind to MMS Proxy-Relay | 7.2.1 | O | MMSE-S-077 |

**Figure 7.1** Partial SCR definition/Source [OMA-MMS-CTR] (version 1.1)

the list of mandatory and optional requirements to be fulfilled by the MMS client and/or the MMSC. An SCR definition also expresses dependencies towards other OMA specifications.

Figure 7.1 shows a partial SCR definition extracted from [OMA-MMS-CTR] (version 1.1). This definition identifies the optional and mandatory server features for the support of the notification transaction.

An SCR definition is presented as a table with the following columns:

- *Item*: identifier for the feature/requirement (e.g. MMSCTR-NTF-S-001).
- *Function*: short description of the feature (e.g. 'Notification Transaction between MMS Proxy-Relay and Receiving MMS Client').
- *Reference*: section number of the OMA specification with more details on the feature (e.g. 7.2).
- *Status*: whether support of the feature is mandatory or optional. 'O' stands for optional, whereas 'M' stands for mandatory.
- *Requirement*: other features required by this feature (e.g. MMSCTR-NTF-S-002 AND MMSCTR-NTF-S-003).

The identifier (item) is constructed according to the convention shown in Figure 7.2.

In the context of MMS, the three OMA normative MMS specifications include SCR definitions [OMA-MMS-CTR; OMA-MMS-Enc; OMA-MMS-Conf]. Being informative, [OMA-MMS-Arch] does not contain any SCR definition.

### 7.1.2 Enabler Implementation Conformance Statement

The Enabler Implementation Conformance Statement (EICS or ICS) is a statement of the capabilities and options supported by a device (MMSC or MMS client) for a given enabler release. It indicates, for an implementation, the set of mandatory and optional features it supports for each specification/standard in an enabler release. The ICS is basically built up from SCR definitions contained in normative documents composing an enabler release. For MMS, OMA published an ICS template for the MMS 1.1 enabler release [OMA-MMS-ICS]. It is usually up to the device vendor to complete this template in order to produce the device ICS.

**Specification name**
MMSCTR is [OMA-MMS-CTR]
MMSE is [OMA-MMS-Enc]
MMSCONF is [OMA-MMS-Conf]

**Device type**
C: (MMS) Client
S: Server (MMSC)

# MMSCTR-NTF-S-001

**Group type**
NTF is for notification
SND is for sending
FTC is for retrieve/fetch
etc.

**Numeric identifier**
A 3-digit identifier,
unique in the group
type.

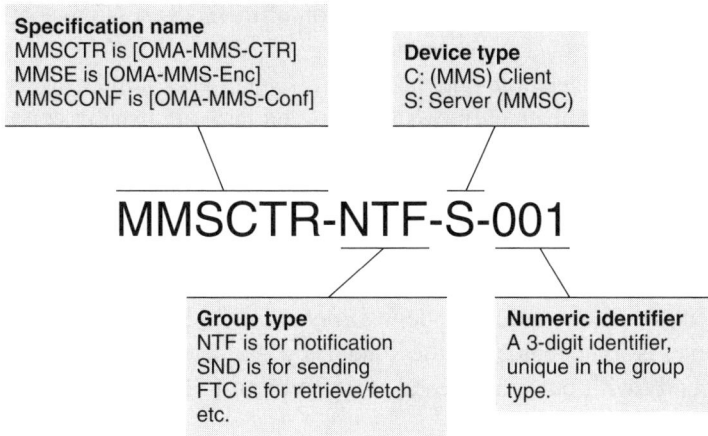

**Figure 7.2**  Convention for building feature/requirement identifiers

### 7.1.3 Enabler Test Requirements, Plan and Specification

As defined in Section 2.7.1, the OMA IOP working group has the responsibility of testing interoperability and solving identified issues. The OMA IOP Multimedia sub-working group is in charge of carrying such activities for MMS. For this purpose, three types of documents are prepared:

- *Enabler Test Requirements (ETR)*: The ETR identifies all requirements important enough to warrant attention from an interoperability perspective and identifies any technical features that should be covered by interoperability testing.
- *Enabler Test Plan (ETP)*: The ETP defines the test strategy and test methodologies for meeting the requirements in the corresponding ETR. The ETP also defines the scope for testing the corresponding enabler release and identifies the high-level requirements for the test tool, if one is needed.
- *Enabler Test Specification (ETS)*: The ETS defines all test cases for the corresponding enabler release, expected inputs and outputs, responses and behaviour for each single test.

Once the ETS has reached a mature stage, then interoperability testing between MMS devices can be organized, as defined in the next section.

### 7.1.4 Interoperability Testing

OMA has selected three methods for conducting interoperability testing:

- *OMA-hosted test fests*: During OMA-hosted test fests, selected device vendors are able to perform one-to-one device testing with other registered MMS devices.

Testing is performed against the latest version of the ETS. A number of test reports provide the outcome of the test sessions.

- *Bilateral testing between manufacturers*: This method is similar to an OMA test fest except that it involves only two vendors and tests are usually conducted in the premises of one of the vendors.
- *Testing by an OMA approved test house*: This method consists of mandating a third party to conduct interoperability tests on behalf of OMA members.

Once interoperability tests have been successfully conducted for a given device, then the device vendor can claim that the device conforms to a given MMS enabler release. This, of course, is a guarantee for the vendor's customers that the device is interoperating efficiently with other devices also conforming to the same enabler release.

## 7.2 Implementations of Different Versions of the MMS Protocol

As shown in Chapter 6, each MMS Protocol Data Unit (PDU) is marked with an MMS version (e.g. version 1.0 or version 1.1). A PDU of higher version may have more parameters than the equivalent PDU marked with a lower version. Furthermore, certain PDU for a given version may not have any equivalent PDU in the protocol of a lower version (e.g. the dedicated PDU for read reports is defined in MMS version 1.1, but is not available in MMS version 1.0). Communicating MMS devices (MMSC and MMS clients) can conform to different versions of MMS (an MMS client that conforms to MMS 1.0 communicating with an MMSC that conforms to MMS 1.1). Consequently, the MMS standard [OMA-MMS-Enc] (from version 1.1) includes a number of rules for ensuring that such heterogeneous MMS devices interoperate effectively.

The MMS version marking a PDU over the MM1 interface (value assigned to the X-MMS-Version parameter) is composed of a major version number and a minor version number as shown in Figure 7.3.

Two scenarios can occur for communicating MMS devices conforming to different versions of the MMS protocol:

1. *Devices conforming to the same major version number, but to a different minor version number*: An MMS device (MMS client or MMSC) can respond to a

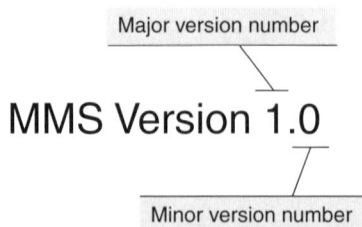

Major version number

# MMS Version 1.0

Minor version number

**Figure 7.3**   Structure of the MMS version

received PDU with a PDU marked with a different minor version number but with the same major version number. An MMS device may, therefore, receive a PDU containing unrecognized parameters. In this context, a receiving MMS client ignores the unrecognized parameters, whereas a receiving MMSC passes unrecognized parameters unchanged (without interpretation). It may happen that the PDU itself is not recognized by the receiving device (unknown value assigned to the X-MMS-Message-Type parameter). In this context, the MMS client responds with a M-notifyresp.ind PDU containing a status value set to 'Unrecognised', whereas the MMSC responds with a M-send.conf PDU with a response value set to 'Error-unsupported-message'.

2. *Devices conforming to the different major version numbers*: Because the behaviour of devices conforming to different major version numbers is expected to be very different, interoperability between such devices is not ensured. In this context, the MMS client that receives a PDU marked with a major version number it does not support responds with a M-notifyresp.ind marked MMS version 1.0 with a status value set to 'Unrecognised'. On the other way round, the MMSC that receives a PDU marked with a major version number it does not support responds with the M-send.conf PDU marked MMS version 1.0 with a response status set to 'Error-unsupported-message'. In the case in which a receiving device supports multiple major version numbers including one of the received PDU, it responds to the received PDU with a PDU marked with the same major version number.

Note that, at the time of writing, all MMS devices available on the market (MMSCs and MMS clients) conformed to the same major version number, which was 1. However, MMS devices conformed to protocol versions with different minor version numbers (e.g. 1.0, 1.1 and 1.2).

# 8

# Commercial Solutions and Developer Tools

Work on the design of standards for the Multimedia Messaging Service (MMS) had been kicked off in 1998. Four years later, first commercial implementations of MMS Centres (MMSCs) and MMS phones appeared on the market. MMS is still in its infancy, but major industry players endorsed the service and have great expectations of its commercial future.

This chapter provides an overview of MMS phones and MMSCs available on the market. It also introduces several tools for application developers.

## 8.1 MMS Handsets Directory

At the time of writing, more than 50 different MMS phone models were available on the market. First MMS phones appeared on the market in the second quarter of 2002, and each new quarter saw the availability of a few more models. Figure 8.1 shows the evolution of the number of available MMS phone models from 2002.

Table 8.1 provides a list of MMS phones available on the market at the time of writing. MMS capabilities indicated in the last column of the table are the ones publicly advertised by vendors in the user agent profile (UAProf, see Section 3.5.4) of each corresponding product. Only retrieving and rendering capabilities are indicated.

## 8.2 MMSC Directory

The following vendors are known to commercialize MMSC solutions: Alcatel, Comverse, Ericsson, Huawei, LogicaCMG, Materna, Nokia, Oksijen Technoloji, Openwave, Symsoft, Tecnomen, TeleDNA, Unisys and Wiral. However, the MMSC market is mainly shared between the following four vendors:

- Nokia
- Ericsson
- LogicaCMG
- Comverse.

*Multimedia Messaging Service: An Engineering Approach to MMS*   Gwenaël Le Bodic
© 2003 John Wiley & Sons, Ltd   ISBN: 0-470-86253-X

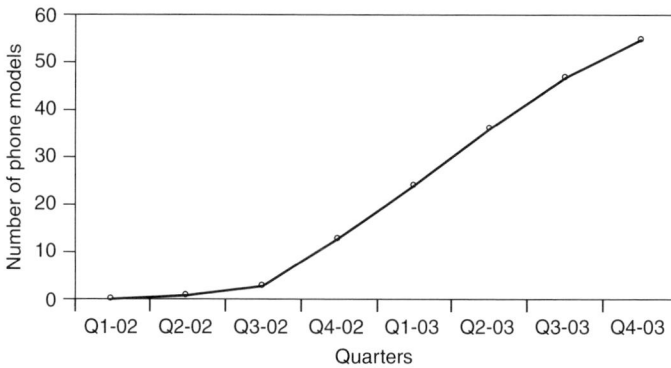

**Figure 8.1**   Availability of MMS phone models (Q1-02 stands for 1st quarter of 2002, Q2-02 stands for 2nd quarter of 2002 and so on.)

## 8.3 Developer Tools

Several companies provide tools to allow the development of MMS software applications. This includes the following tools:

- *Nokia emulator for MMSC interface*: This tool emulates the interface of Nokia's MMSC for Value Added Service (VAS) applications. As described in Chapter 6, this interface is known as the MM7 interface. Note that this tool does not yet support the MM7 interface as described in this book. At the time of writing, the interface between VAS applications and the Nokia MMSC was based on a set of Nokia proprietary commands conveyed over the HyperText Transfer Protocol (HTTP). However, Nokia is expected to provide a version of this emulator supporting the standard MM7 interface as soon as the Nokia's MMSC supports this interface.
- *Nokia MMS Java library*: The Java library can be used for developing MMS-based applications in Java. This library includes the following features:
  — message creation and encoding,
  — message decoding,
  — message sending to Nokia MMSC or to Nokia emulator MMSC.
- *Nokia developer's suite for MMS*: This is a software add-on for Adobe GoLive 6.0. Adobe GoLive, with this add-on, becomes an authoring tool facilitating the generation of multimedia messages with rich multimedia presentations.
- *Nokia Series 60 SDK Symbian OS and MMS extension*: This software development kit allows the development of applications for devices based on the Series 60 platform. This SDK includes a Series 60 emulator enabling the development and testing of applications with a PC.
- *Nokia mobile Internet tool kit*: This tool can be used for testing multimedia messages. It features a ready-to-use application for creating a message from various media objects. Created messages can be provided to one of Nokia's handset simulator (e.g. Nokia 6590 and Nokia 7210 emulators).

**Table 8.1**  Directory of MMS phones

| Phone model | Digital camera | Screen | WAP | Description and receiving capabilities |
|---|---|---|---|---|
| Alcatel OT 535 | None | 128 × 128 4096 colours | 2.0 | MMS 1.0 Max message size 51,200 bytes • Image: resolution 640 × 480 JPEG, GIF, PNG, BMP, WBMP • Audio: MIDI, SP-MIDI and AMR • SMIL support • No support for PIM elements |
| Alcatel OT 735 | Built-in | 128 × 128 4096 colours | 2.0 | MMS 1.0: • Max message size 51,200 bytes • Image: resolution 640 × 480 JPEG, GIF, PNG, BMP, WBMP • Audio: MIDI, SP-MIDI and AMR • SMIL support • No support for PIM elements |
| LG G4050 | Accessory | 65,536 colours | 2.0 | MMS 1.0 |
| LG G5400 | Not known | 128 × 168 65,536 colours | 1.2.1 | MMS 1.0 |
| LG G7100 | Built-in | 65,536 colours | 1.2.1 | MMS 1.0 |
| LG G8000 | Built-in | 176 × 220 65,536 colours | 2.0 | MMS 1.0 |
| Motorola A760 | Built-in | Touchscreen | Not known | MMS 1.0 Email (POP3 + SMTP) |
| Motorola A820 | Accessory | 176 × 220 4096 colours | 2.0 | MMS 1.0 |
| Motorola A835 | Built-in | 176 × 220 65,536 colours | Not known | MMS 1.0 Dual-mode GPRS/UMTS |
| Motorola E390 | Built-in | 176 × 220 65,536 colours | Not known | MMS 1.0 |
| Motorola T720i | Accessory | 120 × 160 4096 colours | Not known | MMS 1.0 |
| Motorola T725 | Accessory | 120 × 160 4096 colours | Not known | MMS 1.0 |
| Motorola V600 | Built-in | 176 × 220 65,536 colours | 2.0 | MMS 1.0 Instant messaging + email (POP3 and SMTP) |
| NEC e525 | Built-in | 162 × 216 65,536 colours | 1.2.1 | MMS 1.0 |
| NEC e606 | Built-in | 132 × 162 65,536 colours | 2.0 | MMS 1.0 |
| NEC e808 | Built-in (dual image + video) | 132 × 162 65,536 colours | Not known | MMS 1.0 |
| Nokia 3100 | Accessory | 128 × 128 4096 colours | Not known | MMS 1.0 |
| Nokia 3300 Series 40 | Not known | 128 × 128 4096 colours | 1.2.1 | MMS 1.0: • Max message size 46,080 bytes • Image: resolution 352 × 288 JPEG, GIF, PNG, BMP, WBMP • Audio: MIDI, SP-MIDI and AMR • No support for SMIL and PIM elements |

*(continued overleaf)*

**Table 8.1**  (*continued*)

| Phone model | Digital camera | Screen | WAP | Description and receiving capabilities |
|---|---|---|---|---|
| Nokia 3510i<br>Series 30 | Accessory | 96 × 65<br>4096 colours | 1.2.1 | MMS 1.0:<br>• Max message size 46,080 bytes<br>• Audio: MIDI, SP-MIDI/no AMR support<br>• No support for SMIL and PIM elements |
| Nokia 3595<br>Series 30 | None | 96 × 65<br>4096 colours | 1.2.1 | MMS + Instant messaging<br>OMA DRM v1.0 (forward lock)<br>MMS 1.0:<br>• Max message size 46,080 bytes<br>• Image: resolution 352 × 288 JPEG, GIF, PNG, BMP, WBMP<br>• Audio: MIDI and SP-MIDI/no AMR support<br>• No support for SMIL and PIM elements |
| Nokia 3650<br>Series 60 | Built-in | 176 × 208<br>4096 colours | 2.0 | MMS 1.0:<br>• Max message size 102,400 bytes<br>• Image: max resolution 640 × 480 JPEG, GIF, PNG, BMP, WBMP<br>• Audio: MIDI, SP-MIDI, AMR, RMF, Basic<br>• Video: 3GPP (H263 + AMR)<br>• PIM: vCard, vCalendar |
| Nokia 5100<br>Series 40 | Accessory | 128 × 128<br>4096 colours | 1.2.1 | MMS 1.0:<br>• Max message size 46,080 bytes<br>• Image: max resolution 352 × 288 JPEG, GIF, PNG, BMP, WBMP<br>• Audio: MIDI, SP-MIDI/no AMR support<br>• No support for SMIL and PIM elements |
| Nokia 6100<br>Series 40 | None | 128 × 128<br>4096 colours | 1.2.1 | MMS 1.0:<br>• Max message size 46,080 bytes<br>• Image: max resolution 352 × 288 JPEG, GIF, PNG, BMP, WBMP<br>• Audio: MIDI, SP-MIDI/no AMR support<br>• No support for SMIL and PIM elements |
| Nokia 6200<br>Series 40 | None | 128 × 128<br>4096 colours | 1.2.1 | MMS 1.0:<br>• Max message size 46,080 bytes<br>• Image: max resolution 352 × 288 JPEG, GIF, PNG, BMP, WBMP<br>• Audio: MIDI, SP-MIDI/no AMR support<br>• No support for SMIL and PIM elements |
| Nokia 6220<br>Series 40 | Built-in | 128 × 128<br>4096 colours | 2.0 | Instant messaging<br>MMS 1.0:<br>• Max message size 105,000 bytes |

**Table 8.1**   (*continued*)

| Phone model | Digital camera | Screen | WAP | Description and receiving capabilities |
|---|---|---|---|---|
| | | | | • Image: resolution 352 × 288 JPEG, GIF, PNG, BMP, WBMP<br>• Audio: MIDI, SP-MIDI and AMR<br>• No support for SMIL<br>• OMA DRM forward lock |
| Nokia 6600 Series 60 | Built-in (dual image + video) | 176 × 208 65,536 colours | Not known | MMS 1.0 + email (SMTP, POP3 and IMAP) |
| Nokia 6610 | Accessory | 128 × 128 4096 colours | 1.2.1 | MMS 1.0:<br>• Max message size 32,768 bytes<br>• Image: resolution 352 × 288 JPEG, GIF, PNG, BMP, WBMP<br>• Audio: MIDI and SP-MIDI<br>• No support for SMIL and PIM elements |
| Nokia 6650 | Built-in | 128 × 160 4096 colours | 1.2.1 | Dual-mode GSM/WCDMA MMS 1.0:<br>• Max message size 100,000 bytes<br>• Image: max resolution 640 × 480 JPEG, GIF, PNG, BMP, WBMP<br>• Audio: MIDI, SP-MIDI, AMR<br>• Video: 3GPP (H263 + AMR)<br>• No support for SMIL and PIM elements |
| Nokia 6800 Series 40 | Accessory | 128 × 128 4096 colours | 1.2.1 | Email (POP3 and IMAP4) MMS 1.0:<br>• Max message size 46,080 bytes<br>• Image: max resolution 352 × 288 JPEG, GIF, PNG, BMP, WBMP<br>• Audio: MIDI, SP-MIDI<br>• No support for SMIL and PIM elements |
| Nokia 7210 Series 40 | Accessory | 128 × 128 4096 colours | 1.2.1 | MMS 1.0:<br>• Max message size 46,080 bytes<br>• Image: max resolution 352 × 288 JPEG, GIF, PNG, BMP, WBMP<br>• Audio: MIDI, SP-MIDI/no AMR support<br>• No support for SMIL and PIM elements |
| Nokia 7250 Series 40 | Built-in | 128 × 128 4096 colours | 1.2.1 | MMS 1.0:<br>• Max message size 46,080 bytes<br>• Image: max resolution 352 × 288 JPEG, GIF, PNG, BMP, WBMP<br>• Audio: MIDI, SP-MIDI/no AMR support<br>• No support for SMIL and PIM elements |
| Nokia 7650 Series 60 | Built-in | 176 × 208 4096 colours | 1.2.1 | MMS 1.0 |

*(continued overleaf)*

**Table 8.1**  (*continued*)

| Phone model | Digital camera | Screen | WAP | Description and receiving capabilities |
|---|---|---|---|---|
| Nokia 8910i Series 30 | None | 96 × 65 4096 colours | 1.2.1 | MMS 1.0: <br>• Max message size 46,080 bytes <br>• Image: max resolution 352 × 288 JPEG, GIF, PNG, BMP, WBMP <br>• No audio support <br>• No support for SMIL and PIM elements |
| Nokia N-Gage Series 60 | None | 176 × 208 4096 colours | 2.0 | MMS 1.0 |
| Nokia observation camera | Built-in | none | n/a | MMS 1.0: able to send messages but not able to receive messages. |
| Orange SPV Microsoft Smartphone | Accessory | 176 × 220 65,536 colours | | Instant messaging MMS 1.0 |
| Panasonic GD87/GD88 | Built-in | 101 × 176 65,536 colours | 1.2.1 | MMS 1.0: <br>• Max message size 50,000 bytes <br>• Image: max resolution 132 × 134 JPEG, GIF, PNG, BMP, WBMP <br>• Audio: MIDI, iMelody, AMR <br>• SMIL support |
| Panasonic X70 | Built-in | 132 × 176 65,536 colours | Not known | MMS 1.0 |
| Sagem My x6 | Built-in | 128 × 160 65,536 colours | 1.2.1 | MMS 1.0 |
| Samsung P400 | Built-in | 128 × 160 65,536 colours | 1.2.1 | MMS 1.0 |
| Samsung P408 | Not known | Not known | | MMS 1.0 |
| Samsung V200 | Built-in | 128 × 160 65,536 colours | 1.2.1 | MMS 1.0 |
| Sharp GX-1 | Built-in | 120 × 160 65,536 colours | 1.2.1 | MMS 1.0: <br>• Max message size 30,000 bytes <br>• Image: max resolution 120 × 160 JPEG, GIF, PNG, BMP, WBMP <br>• Audio: MIDI, iMelody, WAV/no AMR support |
| Sharp GX-10 | Built-in | 120 × 160 65,536 colours | 1.2.1 | MMS 1.0: <br>• Max message size 30,000 bytes <br>• Image: max resolution 120 × 160 JPEG, GIF, PNG, BMP, WBMP <br>• Audio: MIDI, WAV, iMelody/no AMR support |
| Siemens M55 | Accessory | 101 × 80 4096 colours | 1.2 | MMS 1.0: <br>• Max message size 51,200 bytes <br>• Image: resolution 160 × 120 JPEG, GIF, PNG, BMP, WBMP <br>• Audio: MIDI and AMR <br>• PIM: vCard, vCalendar <br>• SMIL support |
| Siemens S55/S56 | Accessory | 101 × 80 256 colours | 1.2.1 | MMS 1.0: <br>• Max message size 51,200 bytes <br>• Image: max resolution 160 × 120 JPEG, GIF, PNG, BMP, WBMP |

**Table 8.1** (*continued*)

| Phone model | Digital camera | Screen | WAP | Description and receiving capabilities |
|---|---|---|---|---|
| Siemens S55/S57 | Accessory | 101 × 80 4096 colours | 1.2 | • Audio: MIDI, AMR<br>• PIM: vCard, vCalendar<br>MMS 1.0:<br>• Max message size 51,200 bytes<br>• Image: resolution 160 × 120 JPEG, GIF, PNG, BMP, WBMP<br>• Audio: MIDI and AMR<br>• PIM: vCard, vCalendar<br>• SMIL support + email client |
| Siemens SL55 | Accessory | 101 × 80 4096 colours | 1.2.1 | MMS 1.0:<br>• Max message size 51,200 bytes<br>• Image: resolution 160 × 120 JPEG, GIF, PNG, BMP, WBMP<br>• Audio: MIDI and AMR<br>• PIM: vCard, vCalendar<br>• SMIL support |
| Siemens SX1 Series 60 | Built-in | 176 × 220 65,536 colours | 2.0 | Instant messaging MMS 1.0 |
| Sony Ericsson T68i | Accessory | 101 × 80 256 colours | 2.0 | MMS 1.0:<br>• Image: max resolution 160 × 120 GIF, JPEG, WBMP<br>• Audio: AMR, iMelody<br>• PIM: vCard, vCalendar |
| Sony Ericsson T310/T316 | Accessory | 101 × 80 256 colours | 1.2.1 | MMS 1.0:<br>• Image: resolution 160 × 120 JPEG, GIF, WBMP<br>• Audio: MIDI, iMelody, AMR<br>• PIM: vCard, vCalendar<br>• SMIL support |
| Sony Ericsson T610/T616/T618 | Built-in (Flash as accessory) | 128 × 160 65,536 colours | 1.2.1 | MMS 1.0:<br>• Image: resolution 160 × 120 JPEG, GIF, WBMP<br>• Audio: MIDI, iMelody, AMR<br>• PIM: vCard, vCalendar<br>• SMIL support |
| Sony Ericsson P800 | Built-in | 208 × 320 (open) 4096 colours | 2.0 | MMS 1.0:<br>• Max message size 204,800 bytes<br>• Image: resolution 640 × 480 JPEG, GIF, PNG, BMP, WBMP<br>• Audio: MIDI, AMR, WAV, RMF, Basic and iMelody<br>• PIM: vCard, vCalendar<br>• SMIL support |
| Sony Ericsson T300/T302 | Accessory | 101 × 80 256 colours | 2.0 | MMS 1.0:<br>• Image: max resolution 160 × 120 JPEG, GIF, WBMP<br>• Audio: MIDI, AMR, iMelody<br>• PIM: vCard, vCalendar<br>• SMIL support |

- *Comverse development environment*: This development environment contains a set of tools including the following:
  — MMS message player (preview multimedia messages on a variety of MMS phones).
  — MMS software development kit (Java libraries for message composition, submission to Comverse MMS centre, message reception from Comverse MMS centre and message parsing) and WAV to AMR converter.
- *Sony Ericsson MMS home studio*: The studio includes the following:
  — MMS message editor (created message can be sent to an Ericsson phone connected to a PC via infrared link, Bluetooth or serial cable).
  — MMS message viewer (preview the message with Ericsson handset simulator). These tools are available from Ericsson mobility world website (see Box 8.1).
- *Ericsson AMR converter*: This basic tool converts WAV files to AMR files.

---

**BOX 8.1   Resources for developers of MMS applications**

Nokia developer Forum at http://www.forum.nokia.com
Ericsson Mobility World at http://www.ericsson.com/mobilityworld
Comverse developer zone at https://developer.comverse.com/dzone/
Access to resources for developers requires prior online registration.

---

# 9

# The Future of MMS

Commercial Multimedia Messaging Service (MMS) solutions can be improved to guarantee a better level of interoperability with the objective to increase the penetration of MMS devices. In the meantime, industry players are preparing the future of MMS by enhancing existing MMS standards for enabling new features. Three main standardization organizations are involved in this process: 3GPP, 3GPP2 and OMA.

This chapter outlines ongoing developments in relevant standardization bodies.

## 9.1 MMS Developments in 3GPP

The Third Generation Partnership Project (3GPP) has long been involved in the development of several stages of MMS. This organization is not only still in charge of enhancing the existing MMS standards but is also looking at other forms of messaging.

### 9.1.1 MMS Release 6

This book provides an in-depth description of three releases/versions of MMS:

- 3GPP Release 99/WAP Forum version 1.0
- 3GPP Release 4/OMA version 1.1
- 3GPP Release 5/OMA version 1.2.

At the time of writing, 3GPP was carrying work on the next release of MMS – MMS Release 6. This release should be frozen in March 2004, and the Open Mobile Alliance (OMA) will derive a new set of technical specifications based on Release 6 (to be published as OMA MMS version 1.3 or OMA MMS version 2.0).

The scope of the ongoing work for 3GPP MMS Release 6 comprises the following items:

- Enhancements of the MM1 interface (between MMS client and MMSC) and consideration of non-WAP-based MM1 interface realizations (e.g. IP-based).
- Enhancements of the MM4 interface (between two MMSCs).

---

*Multimedia Messaging Service: An Engineering Approach to MMS*   Gwenaël Le Bodic
© 2003 John Wiley & Sons, Ltd   ISBN: 0-470-86253-X

- Enhancements of the MM7 interface (between the MMSC and VAS applications).
- Management of MMS-related information in the (U)SIM.
- Security and privacy.
- Digital Rights Management (DRM).
- Enhancements of charging mechanisms (e.g. definition of the MM8 interface between the MMSC and the billing server).
- Improvements of interworking with external messaging systems (e.g. SMS, Internet email, etc.).
- Interfacing of presence services with MMS.
- Enhancements of streaming support.
- Additional addressing modes (e.g. SIP addressing).
- Additional features for the Multimedia Message Box (MMBox) concept.
- New formats and codecs to be supported in the context of MMS.

The 3GPP MMS Release 6 work items are under the responsibility of the following 3GPP working groups (see Section 2.2):

- *3GPP SA1*: For the definition of MMS requirements.
- *3GPP T2*: For architectural aspects, transactions flows and several low-level interface realizations.
- *3GPP SA4*: For format and codec aspects and for the support of streaming.
- *3GPP SA5*: For charging aspects.

### 9.1.2 IMS Messaging

In 2002, 3GPP initiated standardization work on the definition of IP Multimedia Subsystem (IMS) messaging services [3GPP-22.340]. IMS messaging services identify a group of services supported by capabilities of the 3GPP IP Multimedia Subsystem [3GPP-22.228]. IMS messaging services comprise one or more of the following messaging types:

- *Immediate messaging*: A type of messaging by which the message sender expects that messages are perceived by the recipient as being immediately delivered to the recipient device.
- *Session-based messaging*: A type of messaging by which the message sender expects that messages are perceived by the recipient as being immediately delivered to the recipient device. In addition, users are able to join a messaging session (e.g. chat room).
- *Deferred delivery messaging*: A type of messaging by which the message sender expects the message to be delivered as soon as the recipient becomes available.

This type of messaging relies on the store-and-forward paradigm. MMS is regarded as fulfilling the requirements for deferred delivery messaging.

At the time of writing, 3GPP SA1 was involved in the definition of requirements of immediate messaging and session-based messaging in [3GPP-22.340]. Requirements for deferred messaging are not covered in [3GPP-22.340] since they have already been defined in the MMS technical specification [3GPP-22.140].

## 9.2 MMS Development in OMA

Future work in OMA will consist of building the next version of the MMS standards (client transactions, encapsulation protocol and conformance document). This next version of the MMS standards will build up from the OMA MMS 1.2 enabler release and will fulfil the functional requirements defined by 3GPP in the scope of the Release 6 standard series. Additionally, OMA is expected to widen the number of formats and codecs to be supported by MMS devices (e.g. audio codecs for music teasers, vector graphics, etc.).

Work is also ongoing in OMA for the definition of a standard interface between the MMS Centre (MMSC) and a transcoding server.

## 9.3 MMS Developments in 3GPP2

As described in Chapter 2, 3GPP2 is an organization developing standards for CDMA-based technologies. 3GPP2 is currently aiming at designing alternative realizations of interfaces defined in 3GPP and OMA standards. More specifically, 3GPP2 is carrying work on the design of additional realizations of the MM1 interface using the following IP-based technologies:

- *Session Initiation Protocol (SIP)*: SIP [RFC-3261] is an application-layer control (signalling) protocol for creating, modifying and terminating sessions with one or more participants. Such sessions are typically used for Internet telephone calls, multimedia distribution and multimedia conferences, and 3GPP2 is carrying out work for realizing the MM1 interface of MMS with SIP. In this context, SIP can be used for conveying message notifications and small-size messages only. SIP can appropriately be complemented with the Mobile Interactive Mail Access Protocol (IMAP) for the transport of messages.
- *Mobile IMAP (M-IMAP)*: Mobile-IMAP is an adaptation of IMAP4 rev1 [RFC-3501] that provides minimum overhead message access for mobile messaging.

The following set of 3GPP2 MMS technical specifications are available from the 3GPP2 website at http://www.3gpp2.org:

| [3GPP2-S.R0064.0] | S.R0064-0 – multimedia messaging services, 3GPP2, November 2002 |
| [3GPP2-S0016.000] | X.S0016.000 – MMS specification overview, 3GPP2, April 2003 |
| [3GPP2-S0016.200] | X.S0016.200 – MMS stage 2, functional description, 3GPP2, April 2003 |
| [3GPP2-S016.310] | X.S0016.310 – MMS MM1 stage 3 using OMA/WAP, 3GPP2, April 2003 |
| [3GPP2-S016.340] | X.S0016.340 – MMS MM4 stage 3 intercarrier interworking, 3GPP2, April 2003 |
| [3GPP2-S016.370] | X.S0016.370 – MMS MM7 VASP interworking stage 3 specification, 3GPP2, April 2003 |

3GPP, 3GPP2 and OMA are working closely in order to achieve maximum inter-operability between heterogeneous MMS implementations that will be deployed in different parts of the world.

## 9.4 A Bright Future for MMS?

Considering today's standardization workload for MMS, new enabling technologies are expected to emerge for the support of additional use cases. Photo messaging is being closely followed by video messaging, and the future should open the door for the exchange of even more sophisticated MMS contents such as vector graphics and high-fidelity audio clips.

Today, the deployment of MMS remains in the hands of the mobile phone indus-try, but one can foresee MMS as an interesting platform for the design of hybrid solutions such as personal digital assistants, fixed phones[1], digital cameras and video camcorders with built-in MMS capabilities. Today, users can purchase digital camera accessories for their MMS phones. Soon, users might be able to acquire MMS com-munications accessories for their digital cameras and camcorders. Alternatively, the MMS phone and the digital camera/camcorder could appropriately talk to each other through an infrared port or over a Bluetooth connection.

However, the necessary condition for such a wide-scale support of MMS to other desktop and hand-held appliances is for today's solutions to gain wide acceptance from the mass market of mobile users. The future of MMS is consequently being played now...

---

[1] Fixed Line MMS Forum at http://www.fixedlinemms.org

# Appendices

## A  Content Types of Media Objects

Table A.1 provides the list of content types commonly used for characterizing media objects contained in multimedia messages.

**Table A.1**  Content types of media objects

| Format/codec | | Content-type | |
|---|---|---|---|
| 3GP | | `video/3gpp` | |
| AMR | | `audio/amr` | `audio/x-amr` |
| BMP | | `image/x-bmp` | `image/bmp` |
| GIF | | `image/gif` | |
| JPEG | | `image/jpeg` | `image/jpg` |
| MIDI | | `audio/midi` | `audio/x-midi` |
| PNG | | `image/png` | |
| SMIL | | `application/smil` | |
| SP-MIDI | | `audio/sp-midi` | |
| SVG | | `image/svg + xml` | |
| US-ASCII | | `text/plain;`<br>`    charset = us-ascii` | |
| UTF-8 | [RFC-2279] | `text/plain; charset = utf-8` | |
| UTF-16 | | `text/plain; charset = utf-16` | |
| vCalendar | | `text/x-vCalendar` | |
| vCard | | `text/x-vCard` | |
| WBMP | | `image/vnd.wap.wbmp` | |

## B  MM1 Interface – Response Status Codes (X-Mms-Response-Status)

The confirmation Protocol Data Units (PDUs) `M-send.conf`, `M-forward.conf`, `M-Mbox-delete.conf` and `M-Mbox-view.conf` include a status code indicating

*Multimedia Messaging Service: An Engineering Approach to MMS*   Gwenaël Le Bodic
© 2003 John Wiley & Sons, Ltd   ISBN: 0-470-86253-X

**Table B.1**  Status codes/Response status (X-Mms-Response-Status)

| | Name | B.E. | Description | M-send.conf | M-forward.conf | M-Mbox-delete.conf | M-Mbox-view.conf |
|---|---|---|---|---|---|---|---|
| Transient errors | Ok | 0x80 | The request is accepted by the MMSC. | ✓ | ✓ | | ✓ |
| | Error-transient-failure | 0xC0 | The request is valid but the MMSC is unable to process it because of some temporary conditions. | ✓ | ✓ | | ✓ |
| | Error-transient-sending-address-unresolved | 0xC1 | Owing to some temporary conditions, the MMSC is unable to resolve an address specified in the request. | | | | |
| | Error-transient-message-not-found | 0xC2 | Owing to some temporary conditions, the MMSC is unable to retrieve the message. | | | ✓ | ✓ |
| | Error-transient-network-problem | 0xC3 | Owing to some temporary conditions, the MMSC is unable to process the request. | ✓ | ✓ | | ✓ |
| | Error-transient-partial-success | 0xC4 | The MMSC was not able to successfully complete the requested action for all the indicated multimedia messages. | | | | ✓ |
| | Error-permanent-failure | 0xE0 | An unspecified permanent error occurred during the processing of the request by the MMSC. | ✓ | ✓ | | ✓ |
| | Error-permanent-service-denied | 0xE1 | The request is rejected because of service authentication and authorization failure(s). | ✓ | ✓ | | ✓ |
| | Error-permanent-message-format-corrupt | 0xE2 | The request is badly formatted. | ✓ | ✓ | | ✓ |

**Table B.1**  (*continued*)

| | Name | B.E. | Description | M-send.conf | M-forward.conf | M-Mbox-delete.conf | M-Mbox-view.conf |
|---|---|---|---|---|---|---|---|
| Permanent errors | Error-permanent-<br>sending-address-<br>unresolved | 0xE3 | The MMSC is unable to resolve one of the addresses specified in the request. | √ | √ | | |
| | Error-permanent-<br>message-not-found | 0xE4 | The MMSC is unable to retrieve the message. | | √ | | |
| | Error-permanent-<br>content-not-accepted | 0xE5 | The MMSC cannot process the request because of message size, media types or copyright issues. | √ | | | |
| | Error-permanent-reply-<br>charging-<br>limitations-not-met | 0xE6 | The request does not meet the reply charging limitations. | √ | | | |
| | Error-permanent-reply-<br>charging-request-<br>not-accepted | 0xE7 | The MMSC supports reply charging but the request is rejected because of incompatibility with service or user profile configurations. | √ | | | |
| | Error-permanent-reply-<br>charging-forwarding-<br>denied | 0xE8 | The forwarding request is for a message containing reply charging requirements. | | √ | | |
| | Error-permanent-reply-<br>charging-not-<br>supported | 0xE9 | The MMSC does not support reply charging. | √ | | | |
| | Error-permanent-<br>address-hiding-not-<br>supported | 0xEA | The MMSC does not support address hiding. | √ | | | |
| | Error-unsupported-<br>message | 0x88 | Only used for version management. Used in a response to an unknown PDU or PDU with different major version number. | √ | | | √ |

**Table B.2**   Obsolete error codes/Response status

| | Name | B.E. |
|---|---|---|
| | Error-unspecified | 0x81 |
| | Error-service-denied | 0x82 |
| | Error-message-format-corrupt | 0x83 |
| Obsolete errors | Error-sending-address-unresolved | 0x84 |
| | Error-message-not-found | 0x85 |
| | Error-network-problem | 0x86 |
| | Error-content-not-accepted | 0x87 |

the status of the corresponding request PDU (respectively M-send.req, M-forward.req, M-Mbox-delete.req and M-Mbox-view.req). The status can indicate that the request has been accepted by the MMS Centre (MMSC) or has been rejected because of a permanent or transient error. Status codes, as listed in Table B.1, can be assigned to the X-Mms-Response-Status parameter of the confirmation PDU.

For each PDU, only a set of error codes are applicable. The applicability of an error code to a particular PDU is shown in the three last columns of Table B.1.

The set of error codes introduced in MMS 1.0 [WAP-209] is considered as obsolete for MMS implementations from MMS 1.1 [OMA-MMS-Enc] (version 1.1). Obsolete error codes are supported by MMS entities for ensuring backward compatibility. Obsolete error codes are listed in Table B.2.

Error codes for which the binary encoding representation is in the range 0xE0... 0xFF are all considered as permanent errors. Error codes for which the binary encoding representation is in the range 0xC0... 0xDF are all considered as transient errors.

## C  MM1 Interface – Retrieve Status Codes (X-Mms-Retrieve-Status)

The confirmation PDU M-retrieve.conf can include a status code indicating the status of the corresponding request PDU (WSP/HTTP GET.req). The status can indicate that the request has been accepted by the MMSC or has been rejected because of a permanent or transient error. Status codes, as listed in Table C.1, can be assigned to the X-Mms-Retrieve-Status parameter of the confirmation PDU.

Error codes for which the binary encoding representation is in the range 0xE0... 0xFF are all considered as permanent errors. Error codes for which the binary encoding representation is in the range 0xC0... 0xDF are all considered as transient errors.

**Table C.1**  Status codes/Retrieve status (`X-Mms-Retrieve-status`)

| | Name | B.E. | Description |
|---|---|---|---|
| **Transient errors** | Ok | 0x80 | The request is accepted by the MMSC. |
| | Error-transient-failure | 0xC0 | The request is valid but the MMSC is unable to process it because of some temporary conditions. |
| | Error-transient-message-not-found | 0xC1 | Owing to some temporary conditions, the MMSC is unable to retrieve the message. |
| | Error-transient-network-problem | 0xC2 | The MMSC is unable to process the request because of capacity overload. |
| **Permanent errors** | Error-permanent-failure | 0xE0 | An unspecified permanent error occurred during the processing of the request by the MMSC. |
| | Error-permanent-service-denied | 0xE1 | The request is rejected because of service authentication and authorization failure(s). |
| | Error-permanent-message-not-found | 0xE2 | The MMSC is unable to retrieve the message. |
| | Error-permanent-content-unsupported | 0xE3 | The message contains contents that the recipient MMS client cannot handle and the MMSC is not able to perform the appropriate content adaptation. |

## D  MM1 Interface – MMBox Store Status Codes (X-Mms-Store-Status)

The confirmation PDUs `M-send.conf`, `M-forward.conf`, `M-Mbox-store.conf` and `M-Mbox-upload.conf` include a status code indicating the status of the corresponding request PDU. The status can indicate that the request has been accepted by the MMSC or has been rejected because of a permanent or transient error. Status codes, as listed in Table D.1, can be assigned to the `X-Mms-Store-Status` parameter of the confirmation PDU.

For each PDU, all error codes are sometimes not applicable. The applicability of an error code to a particular PDU is shown in the two last columns of Table D.1.

Error codes for which the binary encoding representation is in the range 0xE5... 0xFF are all considered as permanent errors. Error codes for which the binary encoding representation is in the range 0xC2 . . . 0xDF are all considered as transient errors.

**Table D.1**   Status codes/Store status (X-Mms-Store-Status)

|  | Name | B.E. | Description | M-send.Conf | M-forward.conf | M-Mbox-store.conf | M-Mbox-upload.conf |
|---|---|---|---|---|---|---|---|
| **Transient errors** | Ok | 0x80 | The request is accepted by the MMSC. | | | √ | √ |
| | Error-transient-failure | 0xC0 | The request is valid but the MMSC is unable to process it owing to some temporary conditions. | | | √ | √ |
| | Error-transient-network-failure | 0xC1 | The MMSC was not able to handle the corresponding request owing to capacity overload. | | | √ | √ |
| **Permanent errors** | Error-permanent-failure | 0xE0 | An unspecified permanent error occurred during the processing of the request by the MMSC. | | | √ | √ |
| | Error-permanent-service-denied | 0xE1 | The request is rejected because of service authentication and authorization failure(s). | | | √ | √ |
| | Error-permanent-message-format-corrupt | 0xE2 | The request is badly formatted. | | | | √ |
| | Error-permanent-message-not-found | 0xE3 | The MMSC is unable to retrieve the message. | | | | √ |
| | Error-permanent-mmbox-full | 0xE4 | The user's MMBox is full. | | | √ | √ |

## E  MM4 Interface – Request Status Codes (X-Mms-Request-Status-Code)

The responses PDUs MM4_forward.RES, MM4_delivery_report.RES and MM4_read_reply_report.RES (MM4 interface) include a status code indicating the status of the corresponding request PDU. The status can indicate that the request

**Table E.1**  Request status codes (X-Mms-Request-Status-Code)

| Name | Description |
|------|-------------|
| Ok | The corresponding request was accepted without errors. |
| Error-unspecified | An unspecified error occurred during the processing or reception of the corresponding request. |
| Error-service-denied | The corresponding request was rejected because of failure of authentication or authorization of the originating MMSC. |
| Error-message-format-corrupt | An inconsistency with the message format was detected when the corresponding request was parsed. |
| Error-sending-address-unresolved | There was no MMS address (From, To, Cc) in its proper format or none of the addresses belongs to the recipient MMSC. |
| Error-message-not-found | This status code is obsolete. |
| Error-network-problem | The recipient MMSC was not able to accept the corresponding request owing to capacity overload. |
| Error-content-not-accepted | The message content was not accepted because of message size, media type or copyright issues. |
| Error-unsupported-message | The recipient MMSC does not support the corresponding request PDU. |

has been accepted by the MMSC or has been rejected because of some errors. Status codes, as listed in Table E.1, can be assigned to the X-Mms-Request-Status-Code parameter of the response PDU.

## F  MM7 Interface – Status Code and Status Text

All responses over the MM7 interface include a status code (element StatusCode) of the corresponding request PDU and an optional status text (element Status-Text). The status code can indicate that the request has been accepted by the MMSC or has been rejected owing to some errors. Errors have been organized into four classes:

- Success or partial success
- Client errors
- Server errors
- Service errors.

Status codes, as listed in Table F.1, can be assigned to the StatusCode element of response PDUs.

**Table F.1** Status code

| Err. class | Status code | Status text | Description |
|---|---|---|---|
| Success or partial success | 1000 | Success | This code indicates that the request was executed completely. |
| | 1100 | Partial success | This code indicates that the request was executed partially but some parts of the request could not be completed. Lower order digits and the optional Details parameter may indicate what parts of the request were not completed. |
| Client errors | 2000 | Client error | Client made an invalid request. |
| | 2001 | Operation restricted | The request was refused because of lack of permission to execute the command. |
| | 2002 | Address error | The address supplied in the request was not in a recognized format, or the MMSC ascertained that the address was not valid for the network because it was determined not to be serviced by this MMSC. When used as part of a response, and when multiple recipients are specified in the corresponding submission, this status code indicates that at least one address is incorrect. |
| | 2003 | Address not found | The address supplied in the request could not be located by the MMSC. This code is returned when an operation is requested on a previously submitted message and the MMSC cannot find the corresponding message. |
| | 2004 | Multimedia content refused | The server could not parse the MIME content that was attached to the SOAP message or the content size or media type was unacceptable. |
| | 2005 | MessageID not found | This code is returned when an operation is requested on a previously submitted message and the MMSC cannot find the message for the MessageID specified or when the VAS application receives a report concerning a previously submitted message and the MessageID is not recognized. |
| | 2006 | LinkedID not found | This code is returned when a LinkedID parameter was supplied and the MMSC could not find the related message. |

**Table F.1**   (*continued*)

| Err. class | Status code | Status text | Description |
|---|---|---|---|
| Server errors | 2007 | Message format corrupt | A parameter value format is inappropriate or incorrect. |
| | 3000 | Server error | The server failed to fulfil an apparently valid request. |
| | 3001 | Not possible | This code is normally used as a result of a cancel or status query on a message that is no longer available for such operations. The MMSC has recognized the message in question, but it cannot fulfil the request because the message is already delivered or status is no longer available. |
| | 3002 | Message rejected | Server could not complete the requested service. |
| | 3003 | Multiple addresses not supported | The MMSC does not support this operation on multiple recipients. The operation may be resubmitted as multiple single recipient operations. |
| Service errors | 4000 | General service error | The requested service cannot be fulfilled. |
| | 4001 | Improper identification | Identification header of the request does not uniquely identify the client (either the VAS application or MMSC). |
| | 4002 | Unsupported version | The version indicated by the `MM7Version` parameter is not supported. |
| | 4003 | Unsupported operation | The server does not support the request indicated by the `MessageType` parameter in the header of the message. |
| | 4004 | Validation error | The SOAP and XML structures could not be parsed, mandatory fields are missing or the message format is not compatible to the format specified. The `Details` parameter may specify the parsing error that caused this status. |
| | 4005 | Service error | The operation caused a server (either MMSC or VAS application) failure and should not be resent. |
| | 4006 | Service unavailable | This indication may be sent by the server when service is temporarily unavailable, for example, when server is busy |
| | 4007 | Service denied | The client does not have permission or funds to perform the requested operation. |

# References

## Organizations Involved in the Development of Standards and Recommendations

| | |
|---|---|
| 3G Americas | http://www.3gamericas.org/ |
| 3G.IP | http://www.3gip.org/ |
| European Telecommunications Standard Institute (ETSI) | http://www.etsi.org |
| GSM Association (GSMA) | http://www.gsmworld.com/ |
| Global mobile Suppliers Association (GSA) | http://www.gsacom.com/ |
| International Telecommunication Union (ITU) | http://www.itu.org |
| Internet Engineering Task Force | http://www.ietf.org |
| IPv6 Forum | http://www.ipv6forum.com/ |
| Open Mobile Alliance (OMA) | http://www.openmobilealliance.org |
| Third Generation Partnership Project (3GPP) | http://www.3gpp.org |
| Third Generation Partnership Project 2 (3GPP2) | http://www.3gpp2.org |
| UMTS Forum | http://www.umts-forum.org |
| WAP Forum | http://www.openmobilealliance.org |
| World Wide Web Consortium (W3C) | http://www.w3c.org |

## Standards

*3GPP Documents*

[3GPP-01.01]  3GPP TS 01.01, GSM Release 1999 Specifications.
[3GPP-21.101]  3GPP TS 21.101, 3rd Generation Mobile System Release 1999 Specifications.
[3GPP-21.102]  3GPP TS 21.102, 3rd Generation Mobile System Release 4 Specifications.

*Multimedia Messaging Service: An Engineering Approach to MMS*   Gwenaël Le Bodic
© 2003 John Wiley & Sons, Ltd   ISBN: 0-470-86253-X

[3GPP-21.103]    3GPP TS 21.103, 3rd Generation Mobile System Release 5
                 Specifications.
[3GPP-21.905]    3GPP TR 21.905, System Aspects, Vocabulary for 3GPP
                 Specifications.
[3GPP-22.140]    3GPP TS 22.140, Multimedia Messaging Service, Stage 1.
[3GPP-22.228]    3GPP TS 22.228, Service Requirements for the IP Multimedia,
                 Core Network Subsystem, Stage 1.
[3GPP-22.340]    3GPP TS 22.340, IP Multimedia System Messaging, Stage 1.
[3GPP-23.040]    3GPP TS 23.040, Technical Realization of the Short Message
                 Service.
[3GPP-23.140]    3GPP TS 23.140, Multimedia Messaging Service, Functional
                 Description, Stage 2.
[3GPP-26.140]    3GPP TS 26.140, Multimedia Messaging Service, Media Formats
                 and Codecs.
[3GPP-26.233]    3GPP TS 26.233, Transparent End-to-end Packet Switched
                 Streaming Service, General Description.
[3GPP-26.234]    3GPP TS 26.234, Transparent End-to-end Packet-switched
                 Streaming Services Protocols and Codecs.
[3GPP-31.102]    3GPP TS 31.102, Characteristics of the USIM Application.
[3GPP-32.200]    3GPP TS 32.200, Telecommunication Management, Charging
                 Management, Charging Principles.
[3GPP-32.235]    3GPP TS 32.235, Telecommunication Management, Charging
                 Management, Charging Data Description for Application Services.
[3GPP-41.102]    3GPP TS 41.102, GSM Release 4 Specifications.
[3GPP-41.103]    3GPP TS 41.103, GSM Release 5 Specifications.
[3GPP-51.011]    3GPP TS 51.011, Specification of the SIM/Mobile Equipment
                 Interface.

*3GPP2 Documents*

[3GPP2-S.R0064.0]    S.R0064-0, Multimedia Messaging Service, 3GPP2,
                     November 2002.
[3GPP2-S0016.000]    X.S0016.000, MMS Specification Overview, 3GPP2, April
                     2003.
[3GPP2-S0016.200]    X.S0016.200, MMS Stage 2, Functional Description, 3GPP2,
                     April 2003.
[3GPP2-S016.310]     X.S0016.310, MMS MM1 Stage 3 Using OMA/WAP,
                     3GPP2, April 2003.
[3GPP2-S016.340]     X.S0016.340, MMS MM4 Stage 3 Intercarrier Interworking,
                     3GPP2, April 2003.
[3GPP2-S016.370]     X.S0016.370, MMS MM7 VASP Interworking Stage 3
                     Specification, 3GPP2, April 2003.

*ITU Documents*

[ITU-E.164]   ITU-T E.164, The International Public Telecommunication Number Plan, ITU, May 1997.

[ITU-H.263]   ITU-T H.263, Video Coding for Low Bit Rate Communication, ITU, February 1998.

[ITU-I.130]   ITU-T I.130, Method for the Characterization of Telecommunication Services Supported by ISDN and Network Capabilities of an ISDN, ITU, November 1998.


*IETF Documents*

[RFC-791]   Internet Protocol, DARPA Internet Program, Protocol, IETF, September 1981.

[RFC-822]   Standard for the Format of ARPA Internet Text Messages, IETF, August 13, 1982. Note that [RFC-2822] obsoletes [RFC-822].

[RFC-1806]   Communicating Presentation Information in Internet Messages, the Content-Disposition Header, IETF, June 1995.

[RFC-1889]   RTP: A Transport Protocol for Real-time Applications, IETF, January 1996.

[RFC-1893]   Enhanced Mail System Status Codes, IETF, January 1996.

[RFC-2026]   The Internet Standards Process – Revision 3, IETF, October 1996.

[RFC-2045]   Multipurpose Internet Mail Extensions, Part 1: Format of Internet Message Bodies, IETF, November 1996.

[RFC-2046]   Multipurpose Internet Mail Extensions, Part 2: Media Types, IETF, November 1996.

[RFC-2047]   Multipurpose Internet Mail Extensions, Part 3: Message Header Extensions for Non-ASCII Text, IETF, November 1996.

[RFC-2048]   Multipurpose Internet Mail Extensions, Part 4: Registration Procedures, IETF, November 1996.

[RFC-2049]   Multipurpose Internet Mail Extensions, Part 5: Conformance Criteria and Examples, IETF, November 1996.

[RFC-2279]   UTF-8, A Transformation Format of ISO 10646, IETF, January 1998.

[RFC-2327]   Real Time Streaming Protocol (RTSP), IETF, April 1998.

[RFC-2327]   Session Description Protocol, IETF, April 1998.

[RFC-2392]   Content-ID and Message-ID Uniform Resource Locators, IETF, August 1998.

[RFC-2557]   MIME Encapsulation of Aggregate Documents, IETF, March 1999.

[RFC-2617]   HTTP Authentication: Basic and Digest Access Authentication, IETF, June 1999.

[RFC-2821]   Simple Mail Transfer Protocol, IETF, April 2001.

[RFC-2822]   Internet Message Format, IETF, April 2001.
[RFC-2916]   E.164 Number and DNS, IETF, September 2000.
[RFC-3261]   Session Initiation Protocol, IETF, June 2002.
[RFC-3267]   Real-time Transport Protocol Payload Format and File Storage
             Format for the AMR and AMR-WB Audio Codecs, IETF, June 2002.
[RFC-3501]   Internet Message Access Protocol – Version 4 rev1, IETF, March
             2003.


*OMA Documents*

[OMA-ClientProv]   OMA Client Provisioning Enabler Release, including:
                   —Provisioning architecture overview
                   —Provisioning content
                   —Provisioning bootstrap
                   —User agent behaviour
[OMA-DevMan]       OMA Device Management Enabler Release, including:
                   —SyncML device management bootstrap
                   —Device management conformance requirements
                   —Notification initiated session
                   —SyncML device management protocol
                   —SyncML representation protocol device management
                      usage
                   —SyncML device management security
                   —SyncML device management standardized objects
                   —SyncML device management tree and description
[OMA-DRM]          OMA DRM Enabler Release
                   —Digital rights management
[OMA-DRM-CF]       OMA DRM Enabler Release
                   —Content format
[OMA-MMS-Arch]     OMA Multimedia Messaging Service Enabler Release
                   —Architecture overview
[OMA-MMS-Conf]     OMA Multimedia Messaging Service Enabler Release
                   —Conformance document
[OMA-MMS-CTR]      OMA Multimedia Messaging Service Enabler Release
                   —Client transactions
[OMA-MMS-Enc]      OMA Multimedia Messaging Service Enabler Release
                   —Encapsulation protocol
[OMA-MMS-ICS]      OMA Multimedia Messaging Service Enabler Release
                   —Enabler implementation conformance statement
[OMA-UAProf]       OMA User Agent Profile Enabler Release
                   —User agent profile
[OMA-WSP]          Wireless Session Protocol

*WAP Forum Documents*

[WAP-205]    Multimedia Messaging Service, Architecture Overview, WAP Forum, April 2001.
[WAP-206]    Multimedia Messaging Service, Client Transactions, WAP Forum, June 2001.
[WAP-209]    Multimedia Messaging Service, Encapsulation Protocol, WAP Forum, June 2001.
[WAP-219]    TLS Profile and Tunnelling, WAP Forum, April 2001.
[WAP-224]    Wireless Transaction Protocol, WAP Forum, July 2001.
[WAP-237]    Wireless Application Environment Defined Media Type Specification, WAP Forum, May 2001.
[WAP-250]    WAP Push Architectural Overview, WAP Forum, May 2001.
[WAP-277]    XHTML Mobile Profile, WAP Forum, October 2001.


*W3C Documents*

[W3C-HTML4]         (W3C Recommendation) HTML 4.01, W3C, December 1999. http://www.w3.org/TR/html4/
[W3C-PNG]           (W3C Recommendation) Portable Network Graphics 1.0, W3C, October 1996. http://www.w3.org/TR/REC-png.html
[W3C-SMIL]          (W3C Recommendation) Synchronized Multimedia Integration Language 2.0, W3C, August 2001. http://www.w3.org/TR/smil20
[W3C-SOAP]          (W3C note) Simple Object Access Protocol 1.1, W3C, May 2000. http://www.w3.org/TR/SOAP
[W3C-SOAP-ATT]      (W3C note) SOAP Messages with Attachments, W3C, December 2000. http://www.w3.org/TR/SOAP-attachments
[W3C-sRGB]          A Standard Default Color Space for the Internet 1.10, W3C, November 1996. http://www.w3.org/Graphics/Color/sRGB
[W3C-SVG]           (W3C note) Scalable Vector Graphics 1.0, W3C, September 2001. http://www.w3.org/TR/SVG
[W3C-XHTML-Basic]   (W3C recommendation) XHTML Basic Profile, W3C, December 2000. http://www.w3.org/TR/xhtml-basic/


*Other Documents*

[GSMA-IR.34]     IR.34 – Inter-PLMN Backbone Guidelines, GSM Association, March 2003.
[GSMA-IR.52]     IT.52 – MMS Interworking Guidelines, GSM Association, February 2003.
[IANA-MIBEnum]   Character sets – IANA http://www.iana.org/assignments/character-sets

[IMC-vCalendar]     vCalendar – The Electronic Calendaring and Scheduling
                    Format 1.0, Internet Mail Consortium, September 1996.
                    http://www.imc.org/pdi/vcal-10.doc

[IMC-vCard]         vCard – The Electronic Business Card Format 2.1, Internet
                    Mail Consortium, September 1996,
                    http://www.imc.org/pdi/vcard-21.doc

[MMA-MIDI]          The Complete MIDI 1.0 Detailed Specification, Version 96.1,
                    MIDI Manufacturers Association, 1996.

[MMA-SP-MIDI]       Scalable Polyphony MIDI Specification and Device Profiles,
                    version 1.0a, MIDI Manufacturers Association, May 2002.

# Acronyms and Abbreviations

| | |
|---|---|
| 3G | Third Generation |
| 3GPP | Third Generation Partnership Project |
| 3GPP2 | Third Generation Partnership Project 2 |
| AMR | Adaptive Multi-Rate |
| AMR-NB | AMR NarrowBand |
| AMR-WB | AMR WideBand |
| ARCH | ARCHitecture, OMA group |
| ARIB | Association of Radio Industries and Businesses |
| ARPANET | ARPA wide area NETworking |
| ASCII | American Standard Code for Information Interchange |
| ATM | Asynchronous Transfer Mode |
| BIFS | Binary Format for Scenes |
| CC/PP | Composite Capability/Preference Profiles |
| CDMA | Code Division Multiple Access |
| CDR | Charging Data Record |
| CEK | Content Encryption Key |
| CEPT | Conférence Européenne des Postes et Télécommunications |
| CIF | Common Intermediate Format |
| CN | Core Network |
| CO | Cache Operation |
| CPI | Capability and Preference Information |
| CR | Change Request |
| CSD | Circuit Switched Data |
| CWTS | China Wireless Telecommunication Standard |
| DCF | DRM Content Format |
| DID | Document IDentifier |
| DIG | Developer Interest Group, OMA group |
| DM | Device Management, OMA group |
| DNS | Domain Name Server |
| DRM | Digital Rights Management |

| DS | Data Synchronization, OMA group |
| EF | Elementary File |
| EFI | External Functionality Interface |
| EFR | Enhanced Full Rate |
| EICS | Enabler Implementation Conformance Statement |
| EMS | Enhanced Messaging Service |
| ENUM | E.164 Number Mapping |
| ETP | Enabler Test Plan |
| ETR | Enabler Test Requirements |
| ETS | Enabler Test Specification |
| ETSI | European Telecommunications Standard Institute |
| FQDN | Fully Qualified Domain Name |
| GGSN | Gateway GPRS Support Node |
| GIF | Graphics Interchange Format |
| GMT | Greenwich Mean Time |
| GPRS | General Packet Radio Service |
| GRX | GPRS Roaming Exchange |
| GSA | Global mobile Suppliers Association |
| GSM | Global System for Mobile |
| GSMA | GSM Association |
| HLR | Home Location Register |
| HTML | HyperText Markup Language |
| HTTP | HyperText Transfer Protocol |
| HTTPS | HTTP over SSL |
| IAB | Internet Architecture Board |
| IANA | Internet Assigned Numbers Authority |
| ICS | Implementation Conformance Statement |
| IED | Information Element Data |
| IEDL | Information Element Data Length |
| IEI | Information Element Identifier |
| IESG | Internet Engineering Steering Group |
| IETF | Internet Engineering Task Force |
| IMAP | Interactive Mail Access Protocol |
| IMPS | Immediate Messaging and Presence Services |
| IMS | IP Multimedia Subsystem |
| IMSI | International Mobile Subscriber Identity |
| IOP | Interoperability |
| IOT | Interoperability Testing |
| IP | Internet Protocol |
| ISDN | Integrated Services Digital Network |
| ISOC | Internet SOCiety |
| ITU | International Telecommunication Union |

| | |
|---|---|
| JFIF | JPEG File Interchange Format |
| JPEG | Joint Photographic Experts Group |
| LIF | Location Interoperability Forum |
| LOC | LOCation, OMA group |
| MAG | Mobile Application Group, OMA group |
| MAP | Mobile Application Part |
| MCC | Mobile Country Code |
| MCOM | Mobile COMmerce, OMA group |
| MExE | Mobile Execution Environment |
| MGIF | Mobile Games Interoperability Forum |
| MIDI | Musical Instrument Digital Interface |
| M-IMAP | Mobile IMAP |
| MIP | Maximum Instantaneous Polyphony |
| MMA | MIDI Manufacturer Association |
| MMBox | Multimedia Message Box |
| MMS | Multimedia Messaging Service |
| MMSC | MMS Centre |
| MMSE | MMS Environment |
| MMS-IOP | MMS Interoperability group |
| MNC | Mobile Network Code |
| MNP | Mobile Number Portability |
| MPEG | Motion Picture Experts Group |
| MPG | Mobile Protocols Group, OMA group |
| MSISDN | Mobile Station ISDN Number |
| MTA | Mail Transfer Agent |
| MTU | Maximum Transmission Unit |
| MUA | Mail User Agent |
| MWIF | Mobile Wireless Internet Forum |
| MWS | Mobile Web Services, OMA group |
| OMA | Open Mobile Alliance |
| OPS | Operations and Processes, OMA group |
| OTA | Over The Air |
| PAP | Push Access Protocol |
| PCG | Project Coordination Group |
| PCM | Pulse Code Modulation |
| PDA | Personal Digital Assistant |
| PDP | Packet Data Protocol |
| PDU | Protocol Data Unit |
| PIM | Personal Information Manager |
| PKI | Public Key Infrastructure |
| PLMN | Public Land Mobile Network |
| PNG | Portable Network Graphic |

| | |
|---|---|
| PPG | Push Proxy Gateway |
| PSS | Packet-switched Streaming Service |
| QCIF | Quarter CIF |
| QVGA | Quarter VGA |
| RDF | Resource Description Framework |
| REL | RELease planning, OMA group |
| REQ | REQuirement, OMA group |
| RFC | Request For Comments |
| RGB | Red Green Blue |
| RIM | Research In Motion |
| RTP | Real-time Transport Protocol |
| RTSP | Real-Time Streaming Protocol |
| SAR | Segmentation And Reassembly |
| SCD | Specification Change Document |
| SCR | Static Conformance Requirements |
| SDP | Session Description Protocol |
| SEC | SECurity, OMA group |
| SGSN | Serving GPRS Support Node |
| SI | Service Indication |
| SIM | Subscriber Identity Module |
| SIN | Specification Implementation Note |
| SIP | Session Initiation Protocol |
| SL | Service Loading |
| SMF | Standard MIDI File |
| SMIL | Synchronized Multimedia Integration Language |
| SMS | Short Message Service |
| SMSC | SMS Centre |
| SMTP | Simple Mail Transfer Protocol |
| SOAP | Simple Object Access Protocol |
| SP-MIDI | Scalable Polyphony MIDI |
| SSL | Secure Socket Layer |
| SVG | Scalable Vector Graphics |
| SWG | Sub Working Group |
| TCP | Transmission Control Protocol |
| TLS | Transport Layer Security |
| TR | Technical Report |
| TS | Technical Specification |
| TSG | Technical Specification Group |
| TTA | Telecommunications Technology Association |
| TTC | Telecommunications Technology Committee |
| UAProf | User Agent Profile |
| UCS | Universal Character Set |

| UDH | User Data Header |
| UDHL | User Data Header Length |
| UDP | User Datagram Protocol |
| UMTS | Universal Mobile Telecommunications System |
| URI | Uniform Resource Identifier |
| URL | Uniform Resource Locator |
| USIM | UMTS Subscriber Identity Module |
| UTF | UCS Transformation Format |
| VAS | Value Added Service |
| VASP | VAS Provider |
| VGA | Video Graphics Array |
| W3C | World Wide Web Consortium |
| WAE | Wireless Application Environment |
| WAP | Wireless Application Protocol |
| WBMP | Wireless BitMaP |
| WBXML | WAP Binary XML |
| WDP | Wireless Data Protocol |
| WG | Work Group |
| WIM | Wireless Identity Module |
| WML | Wireless Markup Language |
| WP-HTTP | Wireless Profiled HTTP |
| WP-TCP | Wireless Profiled TCP |
| WSP | Wireless Session Protocol |
| WTA | Wireless Telephony Application |
| WTLS | Wireless Transport Layer Security |
| WTP | Wireless Transaction Protocol |
| WV | Wireless Village |
| XHTML | eXtensible HTML |
| XHTML-MP | XHTML Mobile Profile |
| XML | eXtensible Markup Language |

A list of abbreviations and corresponding definitions used in 3GPP specifications is provided in [3GPP-21.905].

# Index